84 Topics in Current Chemistry
Fortschritte der Chemischen Forschung

Bioactive Organo-Silicon Compounds

Springer-Verlag
Berlin Heidelberg GmbH 1979

This series presents critical reviews of the present position and future trends in modern chemical research. It is addressed to all research and industrial chemists who wish to keep abreast of advances in their subject.

As a rule, contributions are specially commissioned. The editors and publishers will, however, always be pleased to receive suggestions and supplementary information. Papers are accepted for "Topics in Current Chemistry" in English.

ISBN 978-3-662-15436-6 ISBN 978-3-540-35244-0 (eBook)
DOI 10.1007/978-3-540-35244-0

Library of Congress Cataloging in Publication Data. Main entry under title: Bioactive organo-silicon compounds. (Topics in current chemistry ; 84) Bibliography: p. "Author index, volumes 26–84": p. Contents: Tacke, R. and Wannagat, U. Syntheses and properties of bioactive organo-silicon compounds. -- Voronkow, M. G. Biological activity of silatranes. 1. Organosilicon compounds -- Physiological effect. 2. Organosilicon compounds. I. Tacke, Reinhold, 1949- II. Wannagat, Ulrich, 1923- III. Voronkov, Mikhail Grigor'evich. IV. Series. QD1.F58 vol. 84 [QP535.S6] 540'.8s [574.1'924] 79-12799

Originally published by Springer-Verlag Berlin Heidelberg New York in 1979.
Softcover reprint of the hardcover 1st edition 1979

Triltsch, Graphischer Betrieb, 8700 Würzburg
2152/3140-543210

Contents

Springer-Verlag	Postfach 105 280 · D-6900 Heidelberg 1 Telephone (0 62 21) 4 87-1 · Telex 04-61 723
	Heidelberger Platz 3 · D-1000 Berlin 33 Telephone (0 30) 82 20 01 · Telex 01-83319
Springer-Verlag New York Inc.	175, Fifth Avenue · New York, NY 10010 Telephone 4 77-82 00

Syntheses and Properties of Bioactive Organo-Silicon Compounds

Reinhold Tacke* and Ulrich Wannagat

Institut für Anorganische Chemie der Technischen Universität, Braunschweig, Germany

Table of Contents

* All correspondence should be addressed to Dr. R. Tacke.

1 Introduction

Silicon is next to oxygen the most abundant element in the lithosphere; the average content amounts about 30% by weight. Inorganic silicon compounds such as silica and silicates form the basis of most of the rocks forming the earth's crust. In the atmosphere, there is no silicon present, except as dust of cosmic and terrestrial origin. The silicon content in the hydrosphere, mainly in form of dissolved silica (silicic acid), is also very small.

But silicon is not only present in the lithosphere, atmosphere, and hydrosphere. Today there is no doubt that silicon compounds play also a significant role in the biosphere. Silicon occurs, at least in trace amounts, in most plant and animal tissues. It plays a particularly important role for many organisms at a lower stage of evolutionary development, such as diatoms, radiolaria, and a few sponges and gastropods, in which an enrichment of silicon in mineral phases of amorphous silica($SiO_2 \cdot nH_2O$) has been found. However, silicon plays also an important role in the organism of higher plants, higher animals, and man, although in the majority of cases the content of this element is relatively small.

So far, most of the biochemical processes involving silicon are unknown. Organosilicon compounds could not be found in nature till now. Whereas the processes of life are able to build up the important C–C and C–H bonds of organic frameworks, no Si–Si, Si–C or Si–H bonds have been detected in living matter so far. However, such compounds can be synthesized in the laboratory; and the question arose: in which way will living organisms react if they are confronted with synthetic organosilicon compounds?

In recent years, scientists have begun conducting systematic experiments to obtain an answer. A great number of organosilicon compounds have been synthesized and tested in different organisms in order to explore the biological properties of these compounds. A new field of research – biochemistry, pharmacology and toxicology of silicon compounds (bio-organosilicon chemistry) – sprang up.

Several reviews[1–11] dealing with the field of "bio-organosilicon chemistry" have appeared in the literature. The most important is the book "Silicon and Life" by M. G. Voronkov et al.[10]. The present review is concentrated on the followings:

a) A short survey of the main lines of research in the chemistry, pharmacology and toxicology of organosilicon compounds.
b) A comprehensive survey on a special area of bio-organosilicon chemistry, which we call "sila-pharmaca". In this field an extensive view is given concerning syntheses and properties of bioactive silicon compounds having analogous structures (sila-substituted) to well known bioactive carbon compounds (= pharmaca). Similarities and differences of the chemical and physical properties, and also the biological activity of analogous carbon and silicon compounds, are discussed.

Several examples are given to demonstrate the synthetic possibilities for the preparation of bioactive silicon compounds.

2 Comparison Between Carbon and Silicon: Physical and Chemical Effects of Sila-Substitution

The given comparison of the chemistry of carbon and silicon only summarizes the most important features. These are taken from more comprehensive surveys[12–22] in order to give a chemical background for bio-organosilicon chemistry.

Both, carbon and silicon are members of Group IV of the Periodic Table. In spite of this close relationship, there are not only similarities but also striking differences between these elements. Carbon and silicon differ in the size of their atoms (covalent radii: $r_C = 77$ pm; $r_{Si} = 117$ pm), their electronegativities (Allred-Rochow: $\chi_C = 2.50$; $\chi_{Si} = 1.74$), and the energies of their outershell electrons (electronic configuration: C: $1s^2 2s^2 p^2$; Si: $1s^2 2s^2 2p^6 3s^2 3p^2$). Because of the position of silicon in the third row of the Periodic Table, the chemistry of this element is influenced by the availability of empty 3d orbitals which are not greatly higher in energy than the silicon 3s and 3p orbitals. The availibility of low lying 3d orbitals to silicon and the possibility of their involvement in bond formation has been used to explain the easy formation of 5- and 6-coordinated silicon complexes, and the unexpected physical properties, stereochemistry, and chemical behaviour of a number of 4-coordinated silicon compounds.

2.1 Physical Properties of Carbon-Element and Silicon-Element Bonds

A comparison of carbon-element bonds (C–El) inside the most important structural frameworks of living matter (and of the most important drugs) with the corresponding silicon-element bonds (Si–El) is given in Table 1.

According to the electronegativities of carbon, silicon and the elements (El) H, N, O, S, F and Cl the polarisation of the covalent C–El bonds is qualitatively the same as that of the corresponding Si–El bonds, with only two exceptions: the polarisation of the Si–H bond ($Si^{\delta\oplus}-H^{\delta\ominus}$) is opposite to that of the C–H bond ($C^{\delta\ominus}-H^{\delta\oplus}$), and the polarisation of the Si–S bond ($Si^{\delta\oplus}-S^{\delta\ominus}$) is opposite to that of the C–S bond ($C^{\delta\ominus}-S^{\delta\oplus}$), which is nearly without polarity. The Si–El bonds always exhibit a higher degree of polarisation than the corresponding C–El bonds. In the Si–C bond system the silicon is the more positive partner ($Si^{\delta\oplus}-C^{\delta\ominus}$).

Comparison of bond energies of C–El bonds and the corresponding Si–El bonds is difficult, because bond dissociation energy data available for silicon compounds are often contradictory: an obvious reflection of the variety of techniques employed, and of the sensitivity of bond energy caused by substituents. Particularly striking are the different values for the Si–O bond, which are refered in literature [e.g. in SiO_2: 111 kcal · mol^{-1}][26]. In Table 1 some average thermochemical bond energies are given, which are suitable for a comparison between analogous Si–El- and C–El bonds. According to these data, Si–El bonds are always stronger than the corresponding C–El bonds, with only two exceptions: Si–H/C–H and Si–S/C–S. While the Si–Si bond is thermodynamically much weaker than the corresponding C–C bond, the Si–C bond exhibits a bond energy similar to that of the C–C bond.

Table 1. Comparison of carbon- and silicon-element bonds

(a)	(b)	(c)	(d)	(e)	(f)
–C–H	– +	0.30	416 (99)	108	109.1 (Paraffinic)
–Si–H	+ –	0.46	323 (77)	146	147.1 [H_2SiF_2]; 148 [$(CH_3)_2SiH_2$]
–C–C–	0	0	356 (85)	154	154.1 (Paraffinic)
–Si–C–	+ –	0.76	301 (72)	188	186 [$(CH_3)_2SiH_2$]; 188 [CH_3SiF_3]
–Si–Si–	0	0	226 (54)	234	232 [Si_2H_6]; 234 [$(CH_3)_3SiSi(CH_3)_3$]
–C–$\overline{N}$–	+ –	0.57	285 (68)	147	147.2 (Paraffinic, 3 co-valent nitrogen)
–Si–$\overline{N}$–	+ –	1.33	335 (80)	180	172 [$(CH_3)_3SiNHCH_3$]; 173.8 [$(H_3Si)_3N$]
–C–$\overline{\underline{O}}$–	+ –	1.00	336 (80)	143	143 (Paraffinic)
–Si–$\overline{\underline{O}}$–	+ –	1.76	368 (88)	177	163 [$(CH_3)_3SiOSi(CH_3)_3$]; 164 [$Si(OCH_3)_4$]
–C–$\overline{\underline{S}}$–	– +	0.06	272 (65)	181	181(5) (Paraffinic)
–Si–$\overline{\underline{S}}$–	+ –	0.70	226 (54)	217	214 [Cl_3SiSH]; 213.6 [$H_3SiSSiH_3$]
–C–$\overline{\underline{F}}$|	+ –	1.60	485 (116)	136	138.1 (Paraffinic, monosubstituted)
–Si–$\overline{\underline{F}}$|	+ –	2.36	582 (139)	170	154 [SiF_4]; 159.3 [H_3SiF]
–C–$\overline{\underline{Cl}}$|	+ –	0.33	327 (78)	176	176.7 (Paraffinic, monosubstituted)
–Si–$\overline{\underline{Cl}}$|	+ –	1.09	391 (93)	209	200 [$SiCl_4$]; 204.8 [H_3SiCl]

a Carbon- and silicon-element single bonds in 4-coordinated carbon and silicon compounds.
b Polarities of the bonds.
c Differences of electronegativity (Allred-Rochow); Ref. [23].
d Average thermochemical bond energies ($KJ \cdot mol^{-1}$), Ref. [24]; in brackets converted values in $kcal \cdot mol^{-1}$.
e Calculated bond distances (pm; according to Schomaker/Stevenson), Ref. [23].
f Experimental bond distances (pm; corresponding compounds in brackets); Ref. [25] for the C–El bonds, Ref. [13] for the Si–El bonds.

Table 2. Comparison of carbon- and silicon-element bonds

(a)	(b)	(c)	(d)	(e)	(f)	(g)	(h)
–C–H	σ	—	—	—	—	—	—
–Si–H	σ	—	—	→Si–H	—	—	—
–C–C–	σ	>C=C<	–C≡C–	—	—	—	—
–Si–C–	σ[a]	—	—	→Si–C–	→→Si–C–	—	—
–Si–Si–	σ	—	—	→Si–Si–	→→Si–Si–	—	—
–C–N̄<	σ	>C=N̄–	–C≡N̄	—	—	—	—
–Si–N̄<	$\sigma + (p{\to}d)_\pi$[b]	—	—	→Si–N̄<	→→Si–N̄<	N→Si–	N→Si←N
–C–Ō–	σ	>C=Ō	–C≡Ō	—	—	—	—
–Si–Ō–	$\sigma + (p{\to}d)_\pi$[b]	—	—	→Si–Ō–	→→Si–Ō–	O→Si–	O→Si←O
–C–F̄|	σ	—	—	—	—	—	—
–Si–F̄|	$\sigma + (p{\to}d)_\pi$[b]	—	—	→Si–F̄|	→→Si–F̄|	F→Si–	F→Si←F
–C–C̄l|	σ	—	—	—	—	—	—
–Si–C̄l|	$\sigma + (p{\to}d)_\pi$[b]	—	—	→Si–C̄l|	→→Si–C̄l|	—	—
–C–S̄–	σ	>C=S̄[d]	—	—	—	—	—
–Si–S̄–	σ[c]	—	—	—	—	—	—

a Partial double bond character of the $(p \to d)_\pi$-type is observed, if the carbon atom is sp^2- or sp-hybridized.

b Partial double bond character of the $(p \to d)_\pi$-type.

c Little $(p \to d)_\pi$-interaction is postulated for a few Si–S compounds.

d Compounds of this type are relatively uncommon (reluctance of sulfur to form double bonds to carbon).

(a) Carbon- and silicon-element single bonds in 4-coordinated compounds with a central carbon resp. silicon atom.
(b) Character of C-El and Si-El single bonds in 4-coordinated compounds with a central carbon resp. silicon atom.
(c) C-El double bonds of the $(p\text{-}p)_\pi$-type. Analogous Si-El bonds are not stable under normal conditions.
(d) C-El triple bonds of the $(p\text{-}p)_\pi$-type. Analogous Si-El bonds are not stable under normal conditions.
(e) Stable structures, containing a Si-El σ-bond and a 5-coordinated silicon atom. Analogous carbon compounds are unknown.
(f) Stable structures, containing a Si-El σ-bond and a 6-coordinated silicon atom. Analogous carbon compounds are unknown.
(g) Stable structures, containing an El → Si donor bond (σ-type) and a 5-coordinated silicon atom. Analogous carbon compounds are unknown.
(h) Stable structures, containing two El → Si donor bonds (σ-type) and a 6-coordinated silicon atom. Analogous carbon compounds are unknown.

The bond lengths of C–El and Si–El bonds can be calculated according to the rule of Schomaker and Stevenson from covalent radii and differences in electronegativity of the bound atoms. Calculated values of that kind are given in Table 1. In the case of Si–H, Si–S, Si–Si and C–El bonds these values are in good agreement with experimentally observed internuclear distances. However, significant discrepancies from the Schomaker-Stevenson equation are found for the Si–N, Si–O, Si–Cl and Si–F bond: the observed internuclear distances (Table 1) are about 5–15 pm shorter than the calculated values. It has been supposed that these additional bond shortenings are due to $(p\rightarrow d)_\pi$-backbonding: the silicon 3d orbitals are involved in compounds of type $\geqslant Si-\overline{El}$, where El is an atom (bound to further atoms or not) having electrons in a p orbital so situated as to be able to overlap with an empty 3d orbital of silicon. The result is a Si–El bond with partial double-bond character [σ-bond + $(p\rightarrow d)_\pi$], which can be described by the following resonance structures [Eq. (1)]:

$$-\overset{|}{\underset{|}{Si}}-\overline{El} \longleftrightarrow \overset{\ominus}{-}\overset{|}{\underset{|}{Si}}=\overset{\oplus}{El} \qquad (1)$$

This shortening of bond length is accompanied by higher bond energies and valence force constants, as expected for σ single bonds.

It should be pointed out that the Si–C bond length is influenced by the hybridization state of the carbon atom. The Si–C distance in vinyl-, ethinyl- and phenylsilanes is shortened as compared with that of corresponding alkylsilanes [examples[13]: H_3SiCH_3 (186.7 pm), $H_3SiCH{=}CH_2$ (185.3 pm), $H_3SiC{\equiv}CH$ (182.6 pm)].

Whereas carbon is able to form stable $(p-p)_\pi$double and triple bonds with carbon itself, nitrogen and oxygen, leading to coordination numbers 2 and 3 at the carbon atom, silicon cannot build up such bonds (compare Table 2). Although the existence of monomeric species such as SiO and SiNH has been established under specific conditions, and $(p-p)_\pi$ bonded intermediates have been postulated in a number of reactions, no stable silicon analogues of alkenes, alkines, aldehydes, ketones, etc. have been isolated so far. In all cases the σ-bonded oligomer or polymer units have been found instead of their potential π-bonded monomer species [e.g. $(-SiH_2-)_x$ instead of $H_2Si{=}SiH_2$, $\left(-\overset{|}{\underset{|}{Si}}H\right)_x$ instead of $HSi{\equiv}SiH$, $\left(-\overset{R}{\overset{|}{\underset{\underset{R}{|}}{Si}}}-\overline{O}-\right)$ instead of $R_2Si{=}\overline{O}$, etc.].

In contrast to carbon, silicon is able to increase its coordination number from 4 to 5 and 6 ($sp^3 \rightarrow sp^3d \rightarrow sp^3d^2$) by one or two additional $(p \rightarrow d)_\sigma$ donor bonds. 5- and 6-coordinated silicon compounds appear often as reaction intermediates. In several cases such compounds are isolable. The known structure frameworks with carbon- and silicon element bonds, which are stable under normal conditions, are represented in Table 2.

Additional to the bond shortening of certain Si–El bonds, and additional to the increase of the coordination number 4 to 5 and 6, the availability of empty 3d orbitals of silicon is used to explain further differences between carbon compounds and their silicon analogues: Trisilylamine [$N(SiH_3)_3$] was found to have a planar Si_3N-skeleton in contrast to the carbon analogue trimethylamine [$N(CH_3)_3$] which is pyramidal. The structure of the silicon compound was rationalized by assuming that the N atom forms σ-bonds in a trigonal plane leaving the lone pair of electrons in a pure p orbital of nitrogen at right angles to this plane. The electron density from this p orbital is donated into the vacant 3d orbitals of the adjacent silicon atoms. Not only the geometry, but also the weak donor ability of the N atom in $N(SiH_3)_3$ is in agreement with the involvement of this lone pair in bonding to silicon. The same arguments are used to explain the enlarged SiOSi angle of disiloxane ($H_3SiOSiCH_3$) as compared with that of dimethylether (H_3COCH_3).

Further evidence on $(p \rightarrow d)_\pi$-bonding has been obtained by studying the behaviour of the pairs R_3COH/R_3SiOH and $(R_3C)_2NH/(R_3Si)_2NH$ as acids. According to current concepts in organic chemistry the acidities of the –OH and –NH groups should be decreased for the silicon compounds as compared with those of the analogous carbon compounds because of the greater electronegativity of the carbon atom. However, silanols and secondary silylamines were found to be more acidic than the C-analogous carbinols and secondary alkylamines, resp. These results can be also explained by additional $(p \rightarrow d)_\pi$ back donation from the elements nitrogen and oxygen to silicon, leading to an increased s-character in the σ-orbitals of nitrogen and oxygen, because of the changing of hybridization at N and O from sp^3 to sp^2.

Additional to Si–N, Si–O and Si-halogen bonds, silicon can also form partial double bonds [$\sigma + (p \rightarrow d)_\pi$] to carbon. However, such π-interactions are only possible, if the carbon atom is a part of an unsaturated group (sp^2- or sp-hybridization), e.g. in phenyl, vinyl and keto groups, etc. This is shown by the following resonance structures of vinylsilanes [Eq. (2)] and α-silylketones [Eq. (3)]:

$$-\overset{|}{\underset{|}{Si}}-\overset{|}{C}{=}C\!< \;\longleftrightarrow\; -{}^{\ominus}\overset{|}{\underset{|}{Si}}{=}\overset{|}{C}-{}^{\oplus}C\!< \qquad (2)$$

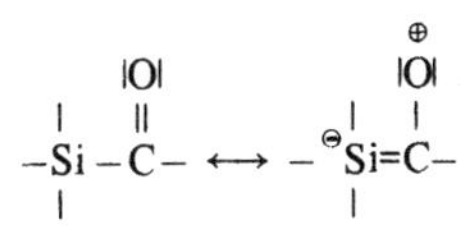

(3)

Spectroscopic data of α-silylketones and the chemical behaviour of vinylsilanes (addition reactions opposite to Markownikoff's rule) are in agreement with this $(p\rightarrow d)_\pi$ model.

Silicon bound to a phenyl group can also influence the bond system by additional $(p\rightarrow d)_\pi$ back donation from carbon to silicon. In agreement with this model, p-trimethylsilyl-substituted benzoic acid shows a greater acidity than expected from inductive effects. Furthermore, p-trimethylsilyl phenol exhibits a greater acidity than phenol itself, and p-trimethylsilyl aniline shows a decreased basicity as compared with that of the nonsubstituted compound. This behaviour can be described by the following resonance structures [Eqs. (4) and (5)]:

(4)

(5)

However, influence on functional groups in organic compounds by a silicon atom is also possible without $(p\rightarrow d)_\pi$ interaction. In examples $(CH_3)_3ElCH_2COOH$ and $(CH_3)_3ElCH_2NH_2$ (El = C, Si) the silicon compound is the weaker acid and the stronger base. These effects are explained by a σ-electron release for the silicon compounds, caused by the lower electronegativity of silicon.

2.2 Chemical Properties of Carbon-Element and Silicon-Element Bonds

Comparison of Si–El and C–El bonds shows not only differences in their physical properties but also in their chemical behaviour. From all C–El and Si–El linkages, described in Table 1, the smallest chemical differences have been found between C–C bond and Si–C bond. The stability of both linkages toward homolytic fission is similar, as can be seen from the small difference between the bond energies (Table 1). However, the heterolytic fission of the Si–C bond usually takes place more readily than that of C–C bond because of the greater ionic character (compare differences of electronegativity, Table 1). The splitting can be achieved by a nucleophilic attack on silicon and by an electrophilic attack on carbon, or by a more or less concerted action of both types of attack. But under normal conditions the Si–C bond is relatively stable to chemical attack, comparable with the C–C linkage. Only in special cases is reactivity of the Si–C bond increased drastically. For example, an exceptional reactivity of β-halogenoalkylsilanes $R_3SiCH_2CH_2X$ (X = halogen) was observed. Compounds of this type of structure are generally much more reactive than the α- or γ-substituted compounds towards heat, basic reagents, and $AlCl_3$. The general reaction involved with all of these reagents is elimination of silyl halide and the formation of olefin [Eqs. (6) and (7)]:

$$(C_2H_5)_3Si{-}CH_2{-}CH_2Cl \xrightarrow{80\,^\circ C} (C_2H_5)_3SiCl + CH_2{=}CH_2 \quad (6)$$

$$Cl(C_2H_5)_2Si{-}CH_2{-}CH_2Cl \xrightarrow{AlCl_3} (C_2H_5)_2SiCl_2 + CH_2{=}CH_2 \quad (7)$$

Other β-functional organosilicon compounds exhibit also abnormal properties. Some examples for this unusual behaviour (denominated by the collective term β-effect)[27] are given by Eqs. (8)–(11):

$$(CH_3)_3Si{-}CH_2{-}CHOH{-}CH_3 \xrightarrow{[H_3O^\oplus]} (CH_3)_3SiOH + CH_2{=}CH{-}CH_3 \quad (8)$$

$$(CH_3)_3Si{-}CH_2{-}CO{-}CH_3 \xrightarrow{C_2H_5OH} (CH_3)_3SiOC_2H_5 + CH_3{-}CO{-}CH_3 \quad (9)$$

$$(CH_3)_3Si{-}CH_2{-}COO{-}C_2H_5 \xrightarrow{HCl} (CH_3)_3SiCl + CH_3{-}COO{-}C_2H_5 \quad (10)$$

$$(CH_3)_3Si{-}CH_2CN \xrightarrow{H_2O} (CH_3)_3SiOH + CH_3CN \quad (11)$$

Very important is the fact that the Si–C bond is stable against hydrolytic attack. Only in special cases (e.g. β-functional substituted organosilicon compounds, $Si{-}C_{aryl}$ bonds in the presence of strong acids) does hydrolysis of the Si–C bond occur.

In contrast to the C–C bond and Si–C bond, the Si–Si linkage has a relative low thermal stability, in good agreement with the low energy of the Si–Si bond (Table 1). However, most reactions of the Si–Si linkage are ionic, mainly consisting of a nucleophilic attack on the silicon atom. Organopolysilanes are relatively stable towards hydrolysis. However, aqueous or alcoholic solutions of alkaline hydroxides cleave the Si–Si bond with liberation of hydrogen [Eq. (12)]:

$$\geq Si{-}Si\leq + 2\,H_2O \xrightarrow{(OH^\ominus)} \geq Si{-}OH + \geq Si{-}OH + H_2 \quad (12)$$

In contrast to this behaviour, hexaalkyldisilanes are relatively resistant towards acids. For example, the reaction of hexamethyldisilane with concentrated H_2SO_4 leads preferentially to a cleavage of the Si–C bond [Eq. (13)]:

$$(CH_3)_3Si{-}Si(CH_3)_3 + 2\,H_2SO_4 \longrightarrow HO{-}SO_2{-}O{-}Si(CH_3)_2{-}Si(CH_3)_2{-}O{-}SO_2{-}OH + 2\,CH_4 \quad (13)$$

However, acids acting on aryl substituted polysilanes lead to a partial cleavage of the Si–Si bond.

In spite of the small difference in the electronegativity of silicon and hydrogen (Table 1), most reactions of the Si–H bond are ionic in character, taking place either by electrophilic attack on hydrogen (which is negatively polarized) or/and by nucleophilic attack on silicon (which is positively polarized). Because of the reverse polarisation the Si–H bond behaves chemically quite differently from the analogous C–H bond. The Si–H linkage can easily be cleaved by water, leading to the corresponding silanols (or disiloxanes) and hydrogen [Eq. (14)]:

$$-\overset{|}{\underset{|}{Si}}-H + H_2O \xrightarrow{(OH^{\ominus})} -\overset{|}{\underset{|}{Si}}-OH + H_2 \qquad (14)$$

This reaction is acid-base catalyzed: alkali catalyzes SiH-hydrolysis more effectively than acid.

Because of the great differences between the electronegativity of silicon and halogens, the silicon-halogen bonds have a high proportion of ionic character. For this reason the Si–F, Si–Cl, and Si–Br bonds are very resistant towards homolytic fission, whereas they are easily cleaved by means of ionic agents, especially by nucleophilic ones. The primary step of most of the reactions based on Si-halogen cleavage seems to be a nucleophilic attack on the silicon atom, which is able to increase its coordination number from 4 to 5 and 6. The reactivity of silicon halides with respect to hydrolytic cleavage reactions decreases in the order Si–I > Si–Br > Si–Cl > Si–F. The reactivity of Si-halogen bonds is drastically increased as compared with that of the analogous C-halogen bonds. For example, CCl_4 is stable against water, while $SiCl_4$ reacts vigorously. Important reasons for this general behaviour are the following: the greater covalent radius of the silicon atom (as compared with carbon) and its capacity, to form a pentavalent (or hexavalent) transition state by increasing its coordination number. These properties of the silicon atom favour the Si-halogen cleavage by a nucleophilic attack of water (or other agents) as compared with the reactivity of analogous C-halogen bonds. The hydrolysis of halogenosilanes (R_3SiX) leads to the corresponding silanols (or disiloxanes) and HX [Eq. (15)]:

$$R_3SiX + H_2O \longrightarrow R_3SiOH + HX \qquad (15)$$

X = halogen

As a result of the great differences in electronegativity between silicon and the elements nitrogen and oxygen, Si–O and Si–N bonds show a high tendency to undergo heterolytic fission, but exhibit a relative great resistance to homolytic cleavage. In contrast to C–O and C–N bonds, analogous Si–O and Si–N linkages show only little resistance towards solvolysis in alcohols and water. Hydrolysis of alkoxysilanes [Eq. (16)] and aminosilanes [Eq. (17)] leads to the corresponding silanols, which can condense under suitable conditions to disiloxanes.

$$R_3SiOR' + H_2O \longrightarrow R_3SiOH + HOR' \qquad (16)$$

$$R_3SiNR'_2 + H_2O \longrightarrow R_3SiOH + HNR'_2 \qquad (17)$$

Such hydrolytic reactions normally take place slowly under neutral conditions.

Hydrolysis of alkoxysilanes is catalyzed by acidic and basic catalysts. Base accelerates the reaction much more efficiently than acids do. But since it proceeds by a nucleophilic attack of the hydroxide ion on silicon, the reaction rate is strongly dependent on the steric environment of the silicon atom and on the electronic effects of its substituents. The acid catalyzed hydrolysis is less sensitive to steric and electronic effects on the silicon atom, because the first step seems to consist in protonation of the alkoxy oxygen.

The hydrolytic cleavage of the Si–N bond of aminosilanes does not proceed so fast as compared to the Si–O and Si–Cl bond in similar alkoxysilanes and chlorosilanes, respectively. However, the hydrolysis can be accelerated by acidic catalysts.

The Si–S bond has some similarities with the Si–O bond. However, its resistance to cleavage by nucleophilic reagents is considerably smaller. In contrast to the C–S bond the Si–S linkage very readily undergoes solvolytic cleavage in water and alcohols. Silthianes are hydrolyzed even by atmospheric moisture, leading to the corresponding silanols.

3 Main Lines of Research in Bio-Organosilicon Chemistry

As a general principle, organosilicon compounds (compounds containing one or more Si–C bonds), although not known in nature, can exhibit biological activity. This statement is the most important result of extensive interdisciplinary research in the novel field of "bio-organosilicon chemistry", carried out by chemists, biologists, pharmacologists, toxicologists, and physicians in the last 10–15 years. Interesting discoveries for this area are to be expected in the future.

Several tests with silicon containing compounds as therapeutics in human medicine have already been crowned with success. In France certain organosilicon preparations, DNR and RDN (compare Chap. 5.5), are used in the treatment of cardiovascular diseases, cancer and virus infections. In the Soviet Union extensive clinical tests with ointments of 1-(chloromethyl)- and 1-ethoxysilatrane as possible drugs for treatment of different typs of alopecia were successful (compare Chap. 5.1). Further clinical studies showed that 1-(chloromethyl)silatrane is also very promising in treatment of wounds and burns. In a Swedish hospital patients with poorly differentiated prostatic carcinoma were treated with 2,6-*cis*-diphenyl-hexamethyl-cyclotetrasiloxane (Cisobitan®): the clinical study also yielded promising results (compare Chap. 5.3).

A large number of organosilicon compounds with high and specific biological activity has been synthesized and investigated pharmacologically and toxicologically in the last years. So far, we can differentiate three main groups of research in the field of biological active organosilicon compounds:

1. Synthesis and biological investigation of silylated derivatives of well known bioactive organic compounds.
2. Synthesis and biological investigation of organosilicon compounds, having no organic analogues at all, or having organic analogues with unknown biological activity.
3. Synthesis and biological investigation of organosilicon compounds, having analogous sila-substituted structures of organic compounds with well known bioactivity (→ sila-pharmaca).

These three main lines of research, which cannot be separated completely from each other in all cases, are characterized by their specific formulation of the problem and by their specific method of investigation. All of them are useful for investigations of the behaviour of silicon compounds in living organisms. Combination of the three main lines is the best approach for the development of organosilicon compounds which can be used therapeutically and prophylactically.

4 Silyl Derivatives of Bioactive Organic Compounds

Most of the silyl derivatives of drugs, described in literature till now, are trimethylsilyl compounds, containing a Si–O, Si–N or Si–C bond between the silicon atom and the drug.

4.1 O- and N-Silylation of Drugs

Silyl derivatives of drugs containing a Si–O or Si–N bond are very easy to prepare. Special silylating agents[28] (such as trimethylchlorosilane, hexamethyldisilazane, etc.) can be used to prepare the corresponding silyl derivatives of drugs, which contain reactive –OH, $-NH_2$ or >NH groups. Some examples[29] for silylation reactions [preparation of N-trimethylsilyl- and N,N-bis-(trimethylsilyl)-derivatives of p-toluenesulphonamide] are given in Eqs. (18)–(20):

(18)

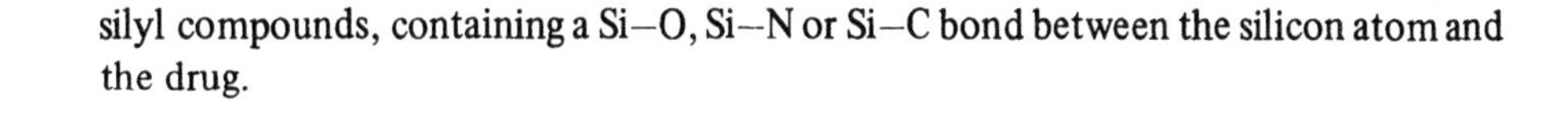

(19)

(20)

The replacement of active hydrogen atoms in drugs by silyl groups leads to a significant alteration of their physico-chemical properties. A very striking effect is the increase of solubility in non-polar solvents. This effect has been shown for several silyl derivatives of drugs, such as the local anaesthetic lidocaine[30], the antiseptic thymol[31], the antipyretic and analgesic paracetamol[32] etc. They exhibit an increased lipophilicity, are readily hydrolyzed by water, and may become interesting objects for pharmacological and toxicological studies, because the altered physicochemical properties have an influence on the pharmacokinetic parameters of these compounds. Blocking of polar centres of a drug by silylation and its subsequent

Table 3. The insect-repellent activity of aminoalkoxysilanes $R[OCH_2CH_2N(C_4H_9)_2]_n$ against X. cheopis[34)]

No.	R	n	Coefficient of repellent action (%) at the dose [g/m^2]			Duration of action (days) at the dose [g/m^2]	
			5	20	40	20	40
2a	H	1	1	60	91	0	0
2b	$(CH_3)_3Si$	1	1	79	98	1	1
2c	$(CH_3)_2Si$	2	99	94	91	12	30
2d	CH_3Si	3	86	95	97	5	12
2e	Si	4	31	68	92	0	0

gradual hydrolysis in the body fluids may offer a simple and interesting route for the development of novel prodrugs, which penetrate easily across lipophilic membranes and liberate the parent drug gradually by hydrolysis.

This behaviour was observed with testosteroxy-trimethylsilane (*1b*) in biological experiments[33)]. The androgenic and myotropic activities of testosterone (*1a*) and its silyl derivative *1b* were estimated by a routine assay (21-day-old castrated male rats; s.c. injection once a day for 7 days; autopsies the day after the last injection). The silyl compound *1b* was more active than testosterone itself. This observation may reflect rapid transport of the silicon compound across the lipid barrier and rapid cleavage to testosterone.

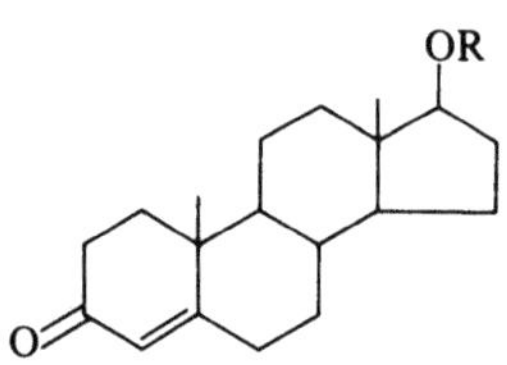

1a: R = H
1b: R = $Si(CH_3)_3$

Lukevics reports[34)] that silylation of 2-dibutylaminoethanol (*2a*) enhances its insect-repellent activity against X. cheopis (Table 3). For silyl derivative *2c* the coefficient of repellent action reaches 99% at a dose of 5 g/m^2. This effect is accompanied by prolongation of the repellent action. According to Lukevics, the activity of silyl derivatives *2b*–*2e* cannot be explained by hydrolysis alone, because this process forms less active compounds. It appears that the repellent properties are concentrated within the molecule of the organosilicon compound itself.

4.2 C-Silylation of Drugs

The preparation of silyl derivatives with a Si–C bond between the silicon atom and the drug is not so easy to realize as compared with Si–O and Si–N systems. In most cases the Si–C bond must be constructed at the beginning of synthesis. This is shown for the triethylsilyl derivative (*4*) of phenethylamine (*3*) in Scheme 1:

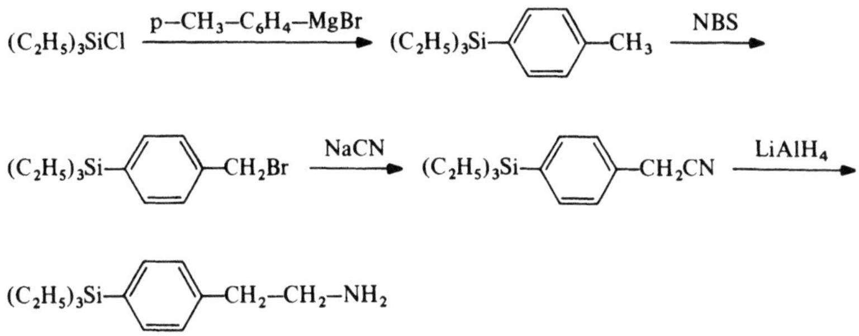

Scheme 1. Synthesis of p-triethylsilylphenethylamine

In contrast to O- and N-silylated compounds, C-silylated drugs are relatively stable against hydrolytic decomposition. For this reason, they have no importance as prodrugs so far.

However, C-silylated drugs can also exhibit remarkable biological activities. For example[35], compounds *4* and *5* have a blood pressure lowering activity (cats and guinea pigs; injection into the left femoral vein), whereas phenethylamine hydrochloride itself exhibits a hypertensive activity. *4* shows a stronger effect than *5*, both on lowering the blood pressure and on the duration of activity. The silyl derivatives *6* and *7* (ortho and meta derivatives of *3*) have a hyperpressure activity similar to phenethylamine.

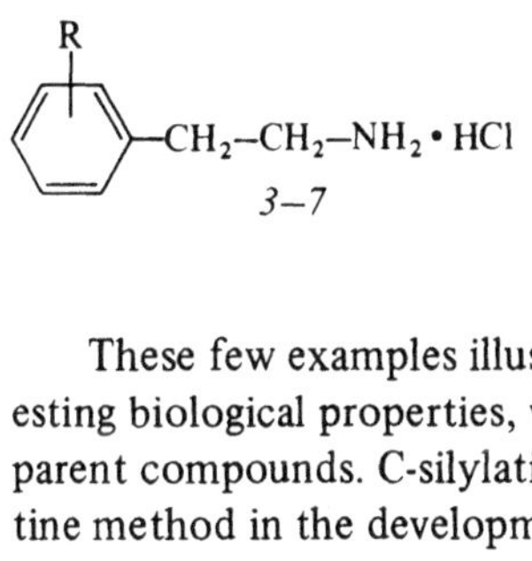

3–7

3: R = H
4: R = p–$Si(C_2H_5)_3$
5: R = p–$Si(CH_3)_3$
6: R = o–$Si(CH_3)_3$
7: R = m–$Si(CH_3)_3$

These few examples illustrate that C-silylation of drugs can also lead to interesting biological properties, which can be similar or different as compared to the parent compounds. C-silylation of bioactive organic compounds may become a routine method in the development of novel drugs.

5 Bioactive Silicon Compounds without Organic Analogues

Investigations of the biological activity of organosilicon compounds which have no organic analogues at all (or which have organic analogues with unknown biological activity) may possibly lead to biotropic agents with novel biological activities. But investigations in this field are very expensive and need luck. They require in the initial step broad screening techniques (bioassays with whole organisms, isolated organs or tissues, population of cells, etc.), to estimate unknown biological properties and to determine the total pharmacological profile of novel organosilicon structures.

However, synthesis and biological screening of hundreds of novel silicon containing compounds resulted in the discovery of a large number of interesting bioactive framework structures, from which only a few characteristic examples are represented in this chapter.

5.1 Silatranes

Silatranes of the general structure *8* (R = alkyl, aryl, alkoxy, aroxy), which have no analogues among organic compounds, represent one of the most fascinating classes of bioactive organosilicon compounds. The theoretical and practical interest is mainly due to their structural features and to their high biological activity.

Most investigations in this field have been carried out by Voronkov and coworkers. Comprehensive surveys[36–38] concerning the chemistry and biological properties of silatranes are given by Voronkov himself. Only the most remarkable and most important results are represented in this chapter in a condensed form.

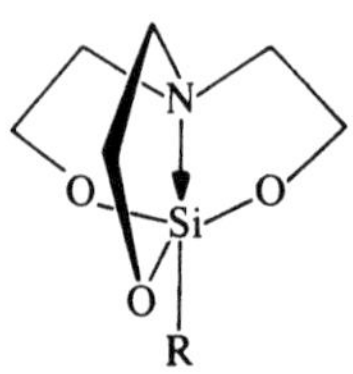

8

Silatranes of type *8* can be described as caged compounds with pentacoordinated silicon and tetracoordinated nitrogen. Physical investigations have given evidence of an intramolecular N→Si coordinative bond. However, a critical analysis[39] of earlier spectroscopic data on silatranes as well as the use of new experimental results (UV-spectroscopy, ^{1}H, ^{13}C, ^{14}N, ^{29}Si NMR, theoretical estimation of dipole moments, etc.) show that the degree of transannular interaction between the silicon and nitrogen atoms was formerly exaggerated. In particular, the electron density transfer from the nitrogen atom to silicon is not so large as was previously considered and does not exceed 0.1–0.2 e.

The first synthesis of silatranes has been described by C. L. Frye et al.[40] and Finestone[41, 42]. In the meantime, further simple and convenient methods for the preparation of silatranes have been found. Some examples [syntheses of 1-phenylsilatrane (*9*), 1-ethoxysilatrane (*18*), and 1-(chloromethyl)silatrane (*15*)] are given in Scheme 2:

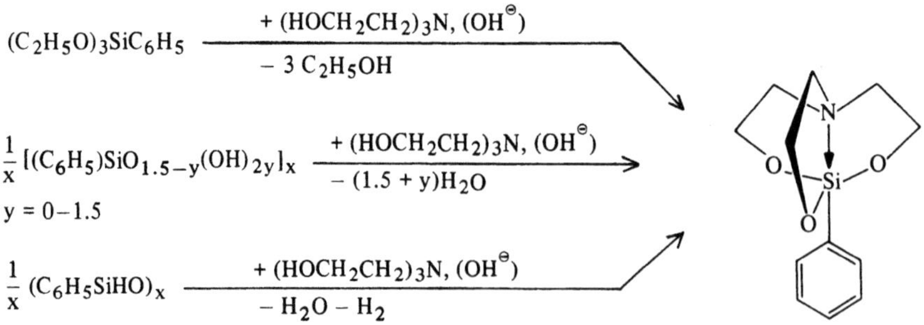

9

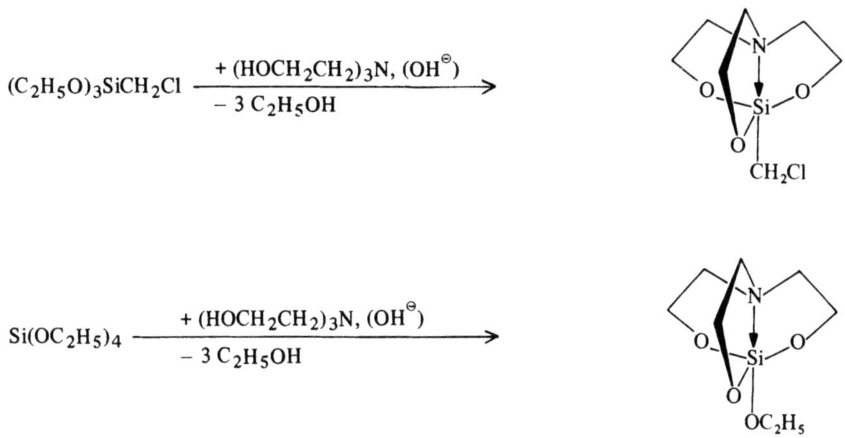

15

18

Scheme 2. Syntheses of silatranes, *9*, *15* and *18*

The most prominent biological feature of silatranes is the remarkable mammalian toxicity exhibited by their 1-aryl derivatives (Table 4). Some of them are several times more toxic than widely known poisons such as hydrocyanic acid and strychnine. At the same time 1-arylsilatranes are almost harmless for cold-blooded animals, plants, and microorganisms. For example, frogs are very resistant to 1-phenylsilatrane (*9*); doses of 30–40 mg/kg have no effect.

Remarkable is the fact, that toxicity of silatranes varies within extremely wide limits, being determined mainly by the nature of the substituents R in structure *8* (Table 4). While 1-arylsilatranes are very toxic, 1-alkyl-, 1-vinyl-, and 1-ethinylsilatranes are practically non-toxic. 1-Alkoxysilatranes have low toxicity and many of them can also be regarded as practically non-toxic.

Comparison of the LD_{50} values of 1-phenylsilatrane (LD_{50} = 0.33 mg/kg), 1-cyclohexylsilatrane (LD_{50}= 150 mg/kg), and 1-benzylsilatrane (LD_{50}=1115 mg/kg) illustrates the strong dependence of toxicity on the substituent R.

Table 4. LD_{50} values[a] of some 1-substituted silatranes (i.p., white mice)

	No.	R	LD_{50} [mg/kg]	No.	R	LD_{50} [mg/kg]
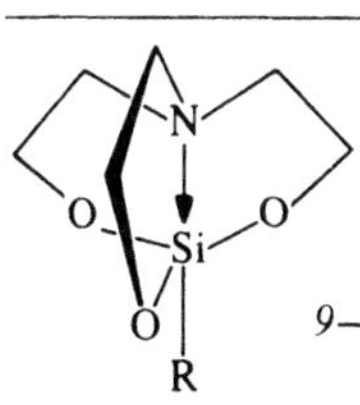 *9–19*	*9:*	C_6H_5	0.33	*15:*	$ClCH_2$	2800
	10:	p-CH_3–C_6H_4	0.20	*16:*	CH_2=CH	3000
	11:	p-Cl–C_6H_4	1.7	*17:*	HC≡C	3000
	12:	cyclo-C_6H_{11}	150	*18:*	C_2H_5O	3000
	13:	C_6H_5–CH_2	1115	*19:*	C_6H_5O	200
	14:	CH_3	3000			

[a] Taken from Ref. [10] for *9*, *10* and *13*;
taken from Ref. [6] for *11*, *14*, *16*, *18* and *19*;
taken from Ref. [38] for *12*, *15* and *17*.

Table 5. Oral toxicity of 1-(p-chlorophenyl)silatrane for warmblooded animals

Species	LD_{50}[mg/kg]
Rats (Norway, laboratory)	1–4
Mice (laboratory)	0.9–2
Sparrows	0.2–0.4
Mallard ducks, pintail ducks	5–10
Monkeys	14.0

The high toxicity of 1-arylsilatranes is already used commercially in the USA. Marketing of 1-(p-chlorophenyl)silatrane (*11*) as a rodenticide began in 1971[43]. It is noted for its high primary toxicity and for its rapid and complete detoxification in the bodies of the poisoned rodents, so that their corpses are harmless for other animals. The oral toxicity of *11*, which has been studied with a limited number of animals, is reported in Table 5.

Numerous other metallatranes have also been tested biologically, but they show much lower effects than the silicon species. The mechanism of the toxic effects of silatranes is not clear in all cases so far.

A further interesting biological feature is the broad spectrum of pharmacological activity in the series of silatranes. For example, 1-phenylsilatrane (*9*) is a powerful analeptic; sublethal doses (0.02–0.25 mg/kg) after intraperitoneal dosing (white mice) lead to aneleptic action as evidenced by motor excitation and accelerated breathing. At a dose of 0.35 mg/kg alternating clonic and tonic convulsions occur. 1-(p-chlorophenyl)silatrane (*11*) and 1-p-tolylsilatrane (*10*) produce also an intensive stimulation of the motor and respiratory centres when administered at lower than lethal doses.

However, 1-(chloromethyl)silatrane (*15*) and 1-ethoxysilatrane (*18*) exhibit other biological properties as compared with those of 1-arylsilatranes. Animal assays showed that both compounds cause complete healing after treatment of wounds. Clinical tests with 1-(chloromethyl)silatrane ointments, carried out in USSR hospitals, confirm these findings. In addition, ointments containing 1-(chloromethyl)-silatrane and 1-ethoxysilatrane lead to an increase of hair growth. Extensive clinical tests with both compounds as possible drugs for the treatment of different types of alopecia have also been carried out in the USSR with positive results.

5.2 Silicon Containing Amines with Insect-Repellent and Antimicrobial Activity

A large number of nitrogen containing organosilicon compounds with insect-repellent and antimicrobial activity is reported by Lukevics[34]. Biological properties of analogous carbon compounds have not been described so far.

Table 6 shows the insect-repellent activity[34, 44] of N,N-dibutylaminomethylsilanes *20–26* against X. cheopis. Compound *22* was found to be the most active agent in this series. At doses of 5 g/m^2, a coefficient of 98% was observed. The longest duration of activity was found for the triethoxy compound *24* (28 days at a dose

Table 6. The insect-repellent activity of N,N-dibutylaminomethyl-silanes $(CH_3)_{3-n}(RO)_nSiCH_2N(C_4H_9)_2$ against X. cheopis[34]

No.	R	n	Coefficient of repellent action [%] at the dose [g/m²]			Duration of action [days] at the dose [g/m²]	
			5	20	40	20	40
20	$(C_4H_9)_2NCH_2CH_2$	1	96	93	93	12	22
21	CH_3CO	1	90	91	95	12	22
22	C_2H_5	1	98	98	100	13	20
23	C_2H_5	2	90	96	98	5	10
24	C_2H_5	3	90	94	75	28	85
25	CH_3	3	65	77	75	28	28
26	$(CH_3)_3Si$	1	90	97	96	5	10

of 20 g/m^2). Several aminoalkyl-substituted disiloxanes also exhibit high insect-repellent activity, for example compound *27*:

$$(C_2H_5)_2N-CH_2-CH_2-CH_2-Si(CH_3)_2-O-Si(CH_3)_2-CH_2-CH_2-CH_2-N(C_2H_5)_2$$

27

Syntheses and biological investigations of about 250 silicon-containing amines led to the discovery of several organosilicon compounds with high antimicrobial activity[34, 45–47]. Some of the most potent agents (*28–32*) are represented in Table 7. Some of them exhibit a fungistatic activity (4–5 μg/ml) against Epidermophyton and Trichophyton which is greater than that of nystatin (7–8 μg/ml). They possess a wider spectrum of action than antifungal antibiotics and also have antibacterial properties.

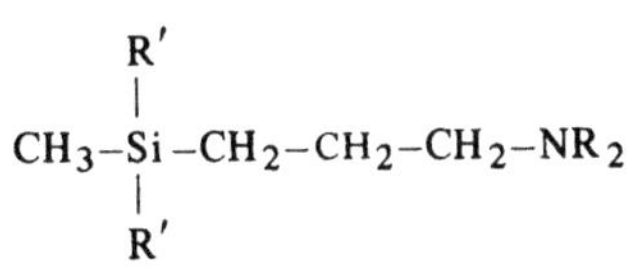

28–32

28: $NR_2 = NC_5H_{10}$; $R' = C_4H_9$
29: $NR_2 = NH_2$; $R' = C_4H_9$
30: $NR_2 = NH_2$; $R' = C_6H_{13}$
31: $NR_2 = NH(C_4H_9)$; $R' = C_4H_9$
32: $NR_2 = N(C_2H_5)_2$; $R' = C_6H_{13}$

Compounds *33* (2.3–4.7 μg/ml) and *34* (1.2–2.3 μg/ml) were also found to have tuberculostatic effects on Mycobacterium tuberculosis[47].

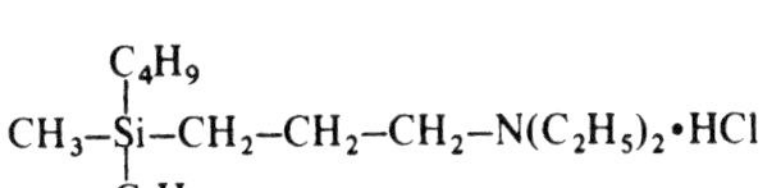

33

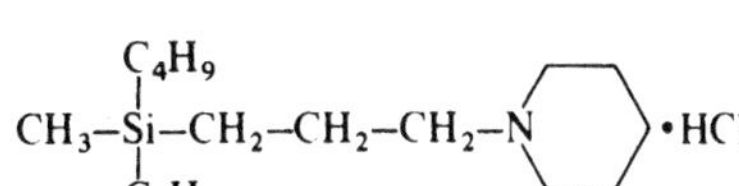

34

Table 7. The minimum concentrations [μg/ml] of (3-aminopropyl)silanes $CH_3R'_2Si(CH_2)_3NR_2$ inhibiting the growth of microorganisms[34)]

No.	Test-microbe Candida albicans 67/846	Epidermophyton Kaufmann-Wolf 41	Tricho-phyton gypseum 4/3	Staphylo-coccus aureus (haemolyticus) 209	Bact. mycoides 537
28	25	12.5	25	2.1	33.3
29	16.7	4.2	8.3	16.6	33.3
30	10.4	10.4	20.9	1.3	2.5
31	10.4	5.2	10.4	0.5	–
32	6.7	5.0	10.4	7.8	–
Nystatin	3.5	6.9	7.8	–	–

5.3 Organosiloxanes

Organopolysiloxanes (silicones) are very important materials for therapeutic applications because of their good physicochemical properties and their inertness to biochemical processes. For example, silicones are employed as ointments (especially for burns), prosthetic materials (e.g. replacement of blood vessels), and plastic surgery (e.g. augmentation of soft tissue or loose skin).

Organopolysiloxanes are also used as ingredients of topical cosmetic formulations, because of their blandness and their capacity to impart water repellency and lubricity to treated surfaces. Compounds, used in this field of application, are mainly low molecular weight organopolysiloxanes.

However, not all organopolysiloxanes are inert to biochemical processes[48–60)]. Toxicological studies on several polysiloxanes resulted in biological effects for a certain experimental cosmetic formulation, which can be described as an equilibrated copolymer of mixed cyclosiloxanes containing at least one phenyl group per cyclic oligomer. The structures concerned have the following general formula *35*:

$$\text{cyclic}[(PhMeSiO)_x(Me_2SiO)_y] \qquad \begin{matrix} Me = CH_3 \\ Ph = C_6H_5 \end{matrix} \qquad x \gtreqless 1;\; x + y = 3 \text{ to } 8$$

35

Subchronic studies[49)] in rabbits treated dermally with *35* resulted in testicular atrophy and spermatogenic depression. The material is also active, when applied by the oral route. Toxicity and reproductive studies in male rats and monkeys showed that *35* is likewise active p.o. in rats and monkeys but inactive in monkeys when administered dermally at daily doses over an extended period of time. Further studies[50)] showed that *35* interrupts the normal oestrous cycle in mature female rats.

In the light of this biological activity, a program was initiated to study various compounds of *35*, which were either isolated from the mixture or synthesized as pure chemical entities. Syntheses and biological investigations of chemically related compounds were also carried out.

As a result of these studies, a series of low molecular weight organosiloxanes, which depressed the reproductive function in the male mouse, rat, and rabbit, were found among phenylmethyl-substituted linear disiloxanes, linear trisiloxanes, cyclic trisiloxanes and cyclic tetrasiloxanes[51, 52]. The spectrum of activity includes testicular atrophy and weight reductions in ventral prostate, seminal vesicle, and vesicular fluid. Cyclic siloxanes are more active than corresponding linear siloxanes. Cyclotetrasiloxane is the most active cycle. Phenyl-heptamethyl-cyclotetrasiloxane (*36*) was found to be a very powerful androgen depressant; however, the presence of an additional phenyl group in the cycle enhances the activity if the two phenyl groups have *cis*-configuration: 2,6-*cis*-diphenyl-hexamethyl-cyclotetrasiloxane (*37*) is the most active compound investigated. The controlled synthesis[61] of the isomers *37* and *38* is described in Scheme 3. X-ray crystallographic investigations[62] revealed that the biologically active *cis*-form has a "boat" conformation in the solid state whereas the trans isomer has a "chair" conformation of the eight-membered tetrasiloxane ring.

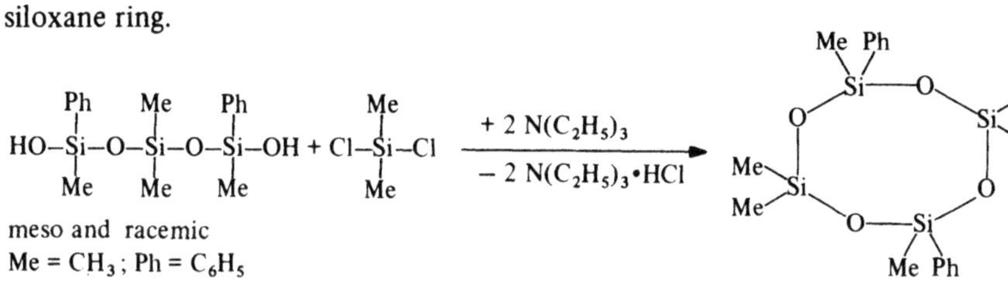

37: *cis*-diphenyl
38: *trans*-diphenyl

Scheme 3. Synthesis of 2,6-*cis*-diphenyl-hexamethyl-cyclotetrasiloxane and its *trans*-isomer

Several low molecular weight organosiloxanes were also examined for effects in the immature female rat[53, 56]. The parameters studied include changes in uterine weight, production of uterine hyperaemia, and uterine histology. Several compounds, which were previously found to be androgenic depressants, showed oestrogen-like activity: phenylmethyl-substituted linear disiloxanes, linear trisiloxanes, cyclotrisiloxanes, and cyclotetrasiloxanes. Cyclotetrasiloxanes as a group possess the greatest potency of any of the compounds tested. Phenyl-heptamethyl-cyclotetrasiloxane and diphenyl-hexamethyl-cyclotetrasiloxane are the most active ones. The biological activity is correlated with high stereospecifity: 2,6-*cis*-diphenyl-hexamethyl-cyclotetrasiloxane (*37*) is the most active compound, being approximately 100 times more potent than the isomeric 2,6-*trans*-diphenyl-hexamethyl-cyclotetrasiloxane (*38*).

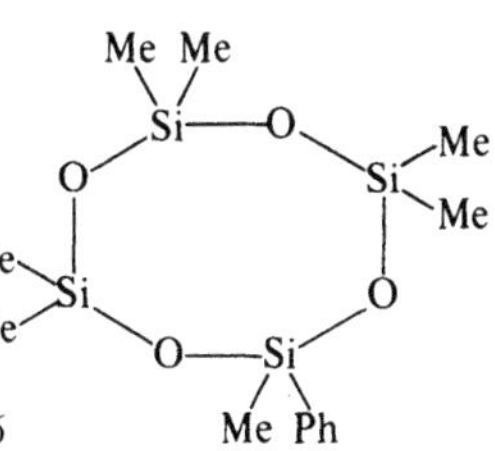

36

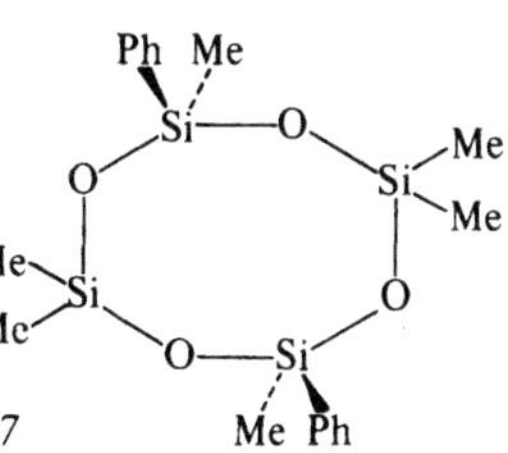

37

(Cisobitan®)

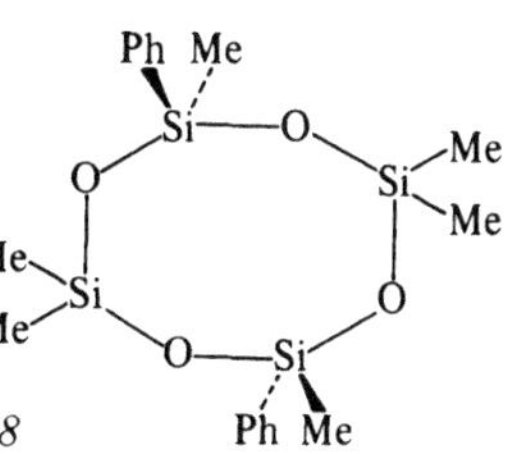

38

Me = CH_3
Ph = C_6H_5

Suppression of the male endocrine system by 2,6-*cis*-diphenyl-hexamethyl-cyclotetrasiloxane was extensively investigated by LeVier[54] and Bennett[56].

37 was shown to have oestrogen-like antigonadotropic activity (orally administered) in the mature male rat, as evidenced by decrease in plasma testosterone, decrease in accessory tissue weight, pituitary enlargement, alterations in pituitary content of follicle-stimulating hormone, luteinizing hormone, and prolactin, and the effects of replacement therapy with androgens and gonadotropins.

LeVier also investigated[55] the hormonal and antifertility activity of 2,6-*cis*-diphenyl-hexamethyl-cyclotetrasiloxane in the female rat. The compound was found to be a perorally active oestrogen in the mature castrate rat: about 1.0 mg/kg is equivalent to 0.005 mg/kg oestradiol benzoate (given subcutaneously) used as a maximum stimulatory dose in an uterotropic assay. *37* is an effective peroral postcoital antifertility agent when administered at 0.33 mg/kg on day 1–5 of gestation and 3.0 mg/kg given on day 1 of gestation. The primary effect appears to be accelerated passage of ova to the uterus and induction of ovum destruction in the oviduct.

2,6-*Cis*-diphenyl-hexamethyl-cyclotetrasiloxane and its *trans*-isomer were also found to increase the brain dopamine content of animals[57]. A dose of 1–100 mg/kg *37* or *38* given daily over a period of about 7 days (orally or intravenously) is effective, depending on the species, the specific mode of administration, the carrier used, and the amount of increase that is desired.

Biochemical effects of the non-steroidal oestrogen 2,6-*cis*-diphenyl-hexamethyl-cyclotetrasiloxane (Cisobitan®) were also investigated in man[58]. Pharmacokinetic and clinical studies have been performed at Radiumhemmet, Stockholm, in patients with poorly differentiated prostatic carcinoma. The pharmaceutical preparation was made as a soy bean oil solution in soft capsules containing 100 mg Cisobitan® per capsule.

In randomized series either Cisobitan®, 100 mg 3 times daily, or Estracyt® (oestramustine phosphate), 300 mg twice daily, was administered orally. As a result after some months (at least 3 months), general condition improved markedly and pain disappeared in the majority of patients in both groups. No drug-related changes in blood chemistry, kidney function tests, haematology or liver enzymes could be detected. An increase in acid and alkaline phosphatase in both groups was found but was more pronounced after giving Cisobitan®. In both groups FSH and LH values were depressed.

Interesting is the fact that during the clinical tests a palliative effect was noted in the form of pain relief[60]. Thus, Cisobitan® and closely related structures were examined for analgesic effects. No activity in mice or rats was found when tested in the hot plate and electro shock models for analgesic activity. However, Cisobitan® was active in a p-benzoquinone-induced stretching model where an increase in stretching rate or hyperalgesia was noted rather than analgesia. This unusual inverse relationship between the effects of Cisobitan® on pain perception in prostatic cancer patients (analgesia) and in laboratory animals (hyperalgesia) appears to be somehow related to oestrogen and/or antigonadotropin activity.

The metabolic fate of Cisobitan® is reported by Vessman et al.[59]. Although the compound is highly lipophilic and nearly water insoluble (solubility: 3 μg/ml), it is eliminated primarily via urine and faeces after peroral administration. Analysis of urinary extracts from the rat, monkey and man revealed the presence of the four metabolic products *39a, 40a, 41a* and *42a,* which were all identified in form of their trimethylsilyl derivatives *39b, 40b, 41b* and *42b,* respectively.

The quantitative analysis was performed with the technique of gas chromatography/mass spectroscopy. A radioactive sample of Cisobitan®, which was labelled with ^{14}C at the geminal methyl groups and which was hexadeuterated in the 4,4-dimethyl-siloxy position, was used for this investigation. Silylation of the metabolic products was carried out with hexamethyldisilazane.

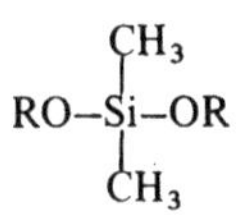

39a: R = H
39b: R = $Si(CH_3)_3$

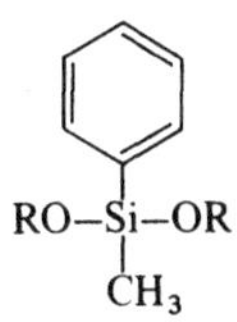

40a: R = H
40b: R = $Si(CH_3)_3$

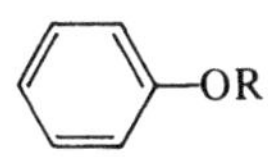

41a: R = H
41b: R = $Si(CH_3)_3$

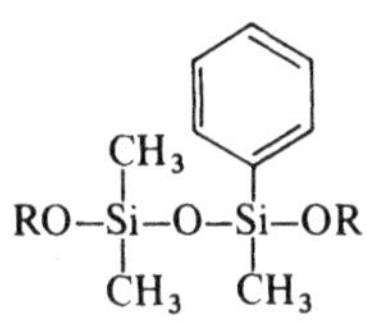

42a: R = H
42b: R = $Si(CH_3)_3$

5.4 Diphenylsilanediol and Related Anticonvulsant Silicon Compounds

According to the rule of Erlenmeyer, compounds of type *43a* with two hydroxyl groups bound to the same carbon atom are not stable and prefer to transform into ketones. The very few organic geminal diols that are isolable are those which carry strongly electron-attracting substituents at the carbon atom.

However, analogous diorganosilanediols *43b* are relatively stable compounds. For example, diphenylsilanediol (*44*), having no carbon analogue, is an isolable crystalline compound, which is easy to prepare by hydrolysis of diphenyldichlorosilane, diphenyldialkoxysilanes, and other diphenyl-substituted hydrolyzable silanes. *44* was synthesized for the first time by Kipping as early as 1912[63].

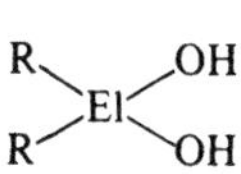

43a: El = C
43b: El = Si

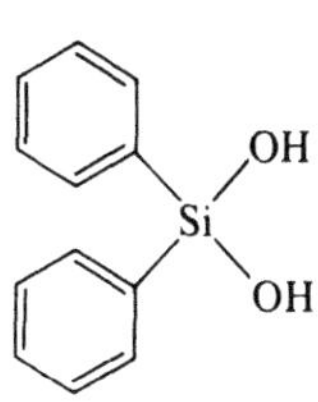

44

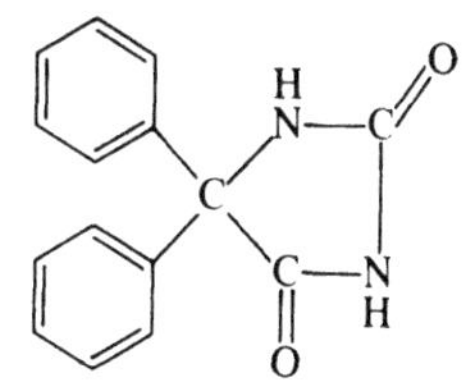

45

Recent reports ascribe anticonvulsant properties to diphenylsilanediol (*44*)[60, 64, 65]. The wide range of action could be demonstrated by its activity against seizures provoked by electroshock and chemoshock: *44* has a peak ED_{50} of 25 mg/kg (given orally) versus maximal electroshock-induced convulsions in mice and rats. The LD_{50} value in mice and rats [intraperitoneal, suspended in methocel] was found to be 2500 mg/kg. Death is preceded by prolonged loss of righting reflex (acute loss at 300 mg/kg). *44* exhibits activity against convulsions provoked by injection of pentylenetetrazol, strychnine or picrotoxin. Its activity is comparable to that of primidone in dogs with idiopathic epilepsy.

The action of *44* appears to be similar to that of the anticonvulsant diphenylhydantoin (*45*). Attempts have been made[66] to correlate the stereochemical and biological features of *44* with those of chemically different anticonvulsants such as diphenylhydantoin, diazepam, procyclidine, trihexyphenidyl, and ethylphenacemide.

Several silicon compounds related to the structure of *44* have also been found to be active in the mouse electroshock test. Some examples are given in Table 8. *44* was the most active agent in this series. It can be seen that increasing as well as decreasing the number of phenyl groups, and hydroxyl groups resp. leads to an decrease of anticonvulsant activity. Compounds *46–52* are capable of hydrolysis in vitro, and presumably in vivo, to give *44*. They appear to be inactive themselves as anticonvulsants but rather act as prodrugs to form the active diphenylsilanediol. Whereas *44* is active immediately after intravenous administration, the non-silanol precursors become active only 15–30 minutes later. Furthermore, activity of these compounds appears to be related to their hydrolysis rate, as can be seen for the dialkoxydiphenylsilanes *47–50*.

Table 8[a]. Effects of silicon compounds *44* and *46–57* in the electroshock test[b]

No.	Compound	ED_{50} for protection [mg/kg] Time after dosing [hours]		
		2	4	6
44	$(C_6H_5)_2Si(OH)_2$	25	37	60
46	$(C_6H_5)_2SiH_2$	28	25	28
47	$(C_6H_5)_2Si(OCH_3)_2$	39	26	52
48	$(C_6H_5)_2Si(OCH_2CH_3)_2$	33	44	53
49	$(C_6H_5)_2Si(OCH_2CH_2CH_3)_2$	>1000	89	32
50	$(C_6H_5)_2Si(OCH_2CH_2CH_2CH_3)_2$	>1000	>1000	>1000
51	$(C_6H_5)_2Si(NH_2)_2$	56	68	100
52	$(C_6H_5)_2Si(OCOCH_3)_2$	34	40	50
53	$C_6H_5Si(OCH_3)_3$	560	300	160
54	$(C_6H_5)_3SiOH$	105	60	110
55	$C_6H_5(CH_3)Si(OH)_2$	80	110	80
56	$(C_6H_5)_2CH_3SiOH$	80	149	80
57	$C_6H_5(CH_3)_2SiOH$	118	135	110

[a] Taken from Ref. [60].
[b] Compounds were given orally in sesame oil at dosages from 10 to 100 mg/kg to male mice.

Remarkable is the fact that diphenylsilanediol and its hydrolyzable prodrugs show a relative wide range between anticonvulsant and neurotoxic dose levels.

Pharmacokinetic studies (e.g. plasma levels in dogs after peroral or intravenous application, tissue distribution in the rat after intravenous administration) with *44* in several animals have been carried out using ^{14}C-labelled diphenylsilanediol and high pressure liquid chromatography. The elimination half-lives were found to be 1.7 h in mice, 7.2 h in rats, and 6.5 h in dogs after peroral application.

Preliminary studies of the metabolism of *44* in dogs resulted in the metabolites *58* and *59*. Additionally, appreciable amounts of diphenylsilanediol could be isolated from both bile and urine. It is not clear so far whether these compounds are excreted as conjugates or not.

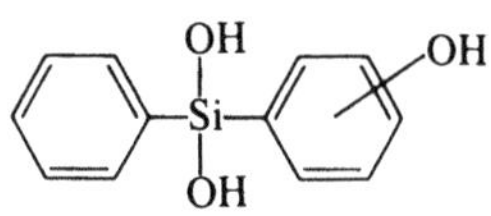

58

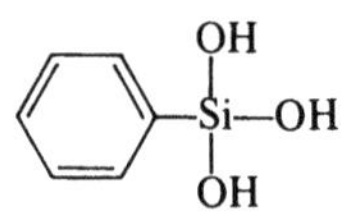

59

5.5 Methylsilanetriol and Dimethylsilanediol

Silicon levels in various organs of man were measured by Loeper et al.[67]. Tendons and aorta possess the greatest content (Fig. 1). A marked decrease in average silicon content in the aorta is observed with age, leading to the hypothetical conclusion that this decrease is an alteration accompanying atherosclerosis.

According to Loeper et al., who have studied the role of silicon in human and experimental atherosclerosis[67], silicon has a protective function for the artery by decreasing the atheromatous deposits and by conserving the integrity of the elastic tissue and the connective tissue.

On this basis a "silicon therapy" was developed in France. Two silicon containing preparations, DNR (*60*) [complex of potassium methylsiliconate with salicylic acid (or with other oxycarbonic acids)] and RDN (*61*) (cyclic ether of dimethylsilanediol and glycerol), are used in the treatment of cardiovascular diseases, cancer and virus infections. Investigations in this field have been carried out by N. Duffaut, J. Loeper, J. Janet, C. Rager, et al. (original literature cited in Ref.[10]).

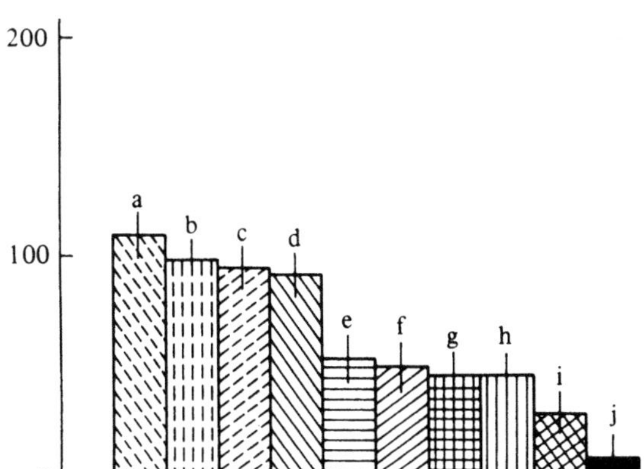

Fig. 1. Silicon levels in the organs of man. (*a*, aorta; *b*, spleen; *c*, tendon; *d*, muscle; *e*, adrenals; *f*, pancreas; *g*, liver; *h*, kidney; *i*, heart; *j*, brain. Ordinate: silicon in μg/100 mg of nitrogen. Bars, mean ± S.E.); taken from Ref.[67]

$CH_3Si(OH)_2OK \cdot n\ RCOOH \cdot mH_2O$

60

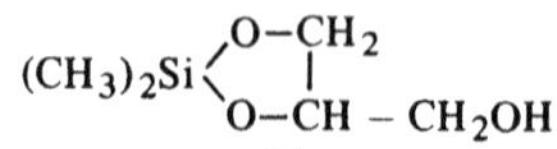

61

Further clinical studies[68] have shown that preparation *62* (i.v., i.m., per os and local administration) is active in painful polycystic mastopathies, probably by regeneration of breast connective tissue. In the course of treatment pain disappeared, the gland became supple again, the nodules diminished in size and there was often massive regression of macrocysts.

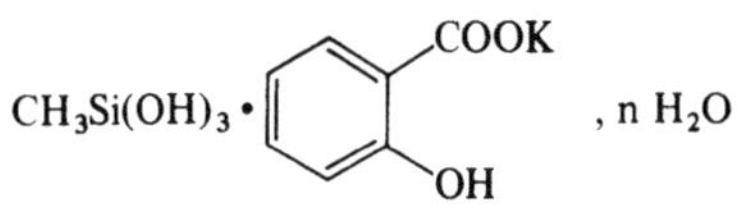

62

6 Sila-Pharmaca

6.1 Comparison of Structures $(\equiv C)_4C$ and $(\equiv C)_4Si$

According to Chap. 2, structures *63a* and *63b* exhibit differences in size, in polarisation of certain bonds (e.g. C–C/Si–C), in basicity (or acidity) of organofunctional substituents (e.g. $\equiv C{-}CH_2{-}\bar{N}\!<$ / $\equiv Si{-}CH_2{-}\bar{N}\!<$) and in chemical reactivity (e.g. α-, β- and γ-effect). Additional effects are possible in the case of aromatic substituents R $[(p \rightarrow d)_\pi$ interactions of the silicon containing systems].

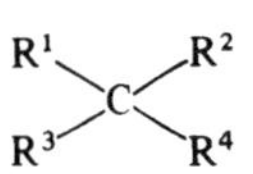

63a

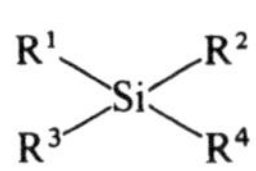

63b

R = aryl and/or alkyl (substituted or not substituted by functional groups)

Differences in physical, chemical and stereochemical properties can be manifested in different bioactivities, if these parameters influence absorption, distribution, mechanism of the molecular action, biotransformation reaction and excretion.

Syntheses and biological tests of several silicon analogues of known bioactive carbon compounds were therefore carried out. The first investigators in this field have been Fregert/Rorsman and Fessenden et al.

6.1.1 Silicon Containing Carbamates with Muscle Relaxant Activity

An extensive study of Fessenden[69] is concerned with biological effects of twelve pairs of carbamates (*64a/64b–75a/75b*) (Table 9), each pair differing by only one atom (silicon in place of carbon). The most prominent of this series is the centrally acting muscle relaxant and tranquilizer meprobamate (*66a*).

Syntheses of *66a* and its silicon analogue sila-meprobamate (*66b*) are described in Scheme 4.

Table 9. Pairs of carbamates *64a/64b–75a/75b* and silicon containing carbamates 76–78

Compound	No.	
$(CH_3)_2El(CH_2OCONH_2)_2$	*64*	
$CH_3(C_2H_5)El(CH_2OCONH_2)_2$	*65*	
$CH_3(n\text{-}C_3H_7)El(CH_2OCONH_2)_2$	*66*	
$CH_3(n\text{-}C_4H_9)El(CH_2OCONH_2)_2$	*67*	
$(CH_3)_2El(CH_2OCONH_2)CH_2CH_2CH_2OCONH_2$	*68*	a: El = C
$(CH_3)_3ElCH_2OCONH_2$	*69*	b: El = Si
$C_2H_5(CH_3)_2ElCH_2OCONH_2$	*70*	
$n\text{-}C_3H_7(CH_3)_2ElCH_2OCONH_2$	*71*	
$n\text{-}C_4H_9(CH_3)_2ElCH_2OCONH_2$	*72*	
$(CH_3)_3El(CH_2)_2OCONH_2$	*73*	
$(CH_3)_3El(CH_2)_3OCONH_2$	*74*	
$(CH_3)_3El(CH_2)_4OCONH_2$	*75*	
$C_6H_5(CH_3)_2SiCH_2OCONH_2$	*76*	
$C_6H_5CH_2(CH_3)_2SiCH_2OCONH_2$	*77*	
$C_6H_5CH_2CH_2(CH_3)_2SiCH_2OCONH_2$	*78*	

The carbamates were tested (intraperitoneal application) using mice for acute toxicity. For sublethal activity the rotating rod test and the extension of hexobarbital sleeping time was used.

Each pair of compounds tested, with one exception, was essentially equivalent in their acute toxicities (compare Table 10). *73b* (450 mg/kg) and *73a* (32 mg/kg) differed by a factor of ~10 in toxicity.

Table 10. Toxicological and pharmacological properties[69, 70] of *64a/64b–75a/75b* and *76–78*

Compound	LD_{50}[a] [mg/kg]	ED_{50}[b] [mg/kg]	Compound	LD_{50}[a] [mg/kg]	ED_{50}[b] [mg/kg]
64a	>1000	>500	*71a*	580	81
64b	>1000	>500	*71b*	400	70
65a	>1000	~900	*72a*	630	135
65b	>1000	339	*72b*	560	110
66a	700	176	*73a*	32	c
66b	>1000	158	*73b*	450	69
67a	>1000	215	*74a*	670	120
67b	900	203	*74b*	470	61
68a	>1000	>630	*75a*	600	139
68b	>1000	238	*75b*	420	118
69a	560	92	*76*	400	111
69b	400	92	*77*	>1000	318
70a	530	81			
70b	600	59	*78*	>1000	308

[a] Calculated after 48 hr after administration (white mice, i.p.).
[b] Rotating rod test (white mice, i.p.).
[c] No activity was noted at sublethal levels.

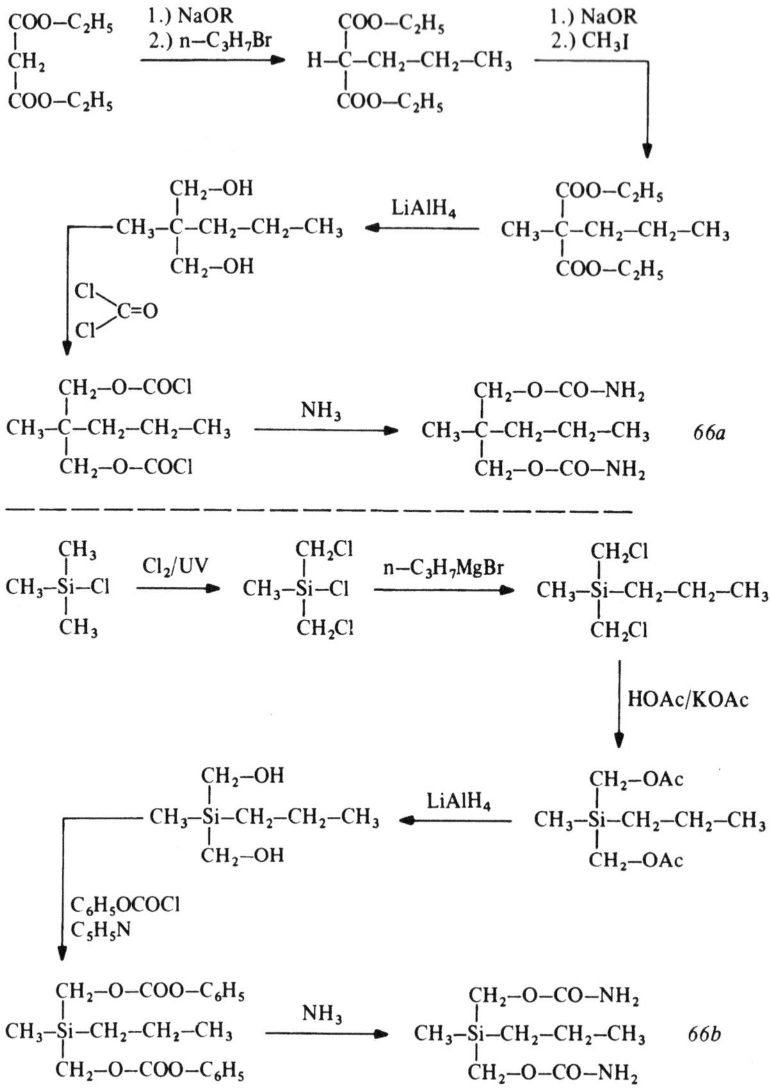

Scheme 4. Syntheses of meprobamate and sila-meprobamate

All compounds, with the exception of *64a*, *64b* and *65a*, exhibited sublethal activities in the rotating rod test. The phenyl-substituted derivatives *76*, *77* and *78* also showed muscle relaxant activity of a short duration[70] when tested by the same animal model (cf. Table 10).

Comparative studies of the pairs *64a*/*64b*–*75a*/*75b* resulted in similar pharmacological effects for the corresponding silicon and carbon compounds, with only a few exceptions: approximately 45 minutes after dosing with *73a*, the animals became convulsive, while animals dosed with *73b* showed only ataxia lasting for approximately 10 minutes. *65a* and *65b* exhibited a significant difference in sublethal activity too; the carbon compound (*65a*) showed no activity in the rotating rod test, while the silicon analogue (*65b*) exhibited a measurable effect. *68a* and *68b* showed also great differences in their activities. However, the other pairs of compounds tested, were essentially equivalent in their sublethal activities or showed only small significant differences.

Remarkable is the fact that *73a* and *73b*, showing the greatest difference in sublethal activity and toxicity, exhibit the same action under invitro conditions. Both compounds were found to be antagonists of the muscarinic activity of acetylcholine and carbachol[71]. The dose-response curves of acetylcholine antagonized by *73a* and *73b* were identical within experimental error.

Very interesting is the comparison between sila-meprobamate (*66b*) and meprobamate (*66a*). Using i.p. application, these compounds do not differ in their effective doses but show a different duration of activity: the carbon compound acts four times longer than its silicon analogue. This pair also differs, when administered orally; in contrast to meprobamate, sila-meprobamate exhibits no appreciable activity (rotating rod test).

This difference could be due either to a lack of absorption of sila-meprobamate when given orally or to a different detoxification pathway. Fessenden und Ahlfors, who investigated the metabolic fate[72] of *64b, 66b* and *71b* after peroral application in rats, could show that sila-meprobamate is absorbed from the gastrointestinal tract. After oral application of *66b*, 60–90% of the ingested silicon was found in the urine within 3 days. On the basis of spectral evidence, Fessenden postulated the structures *79* and *80* for the isolated metabolic products (Scheme 5):

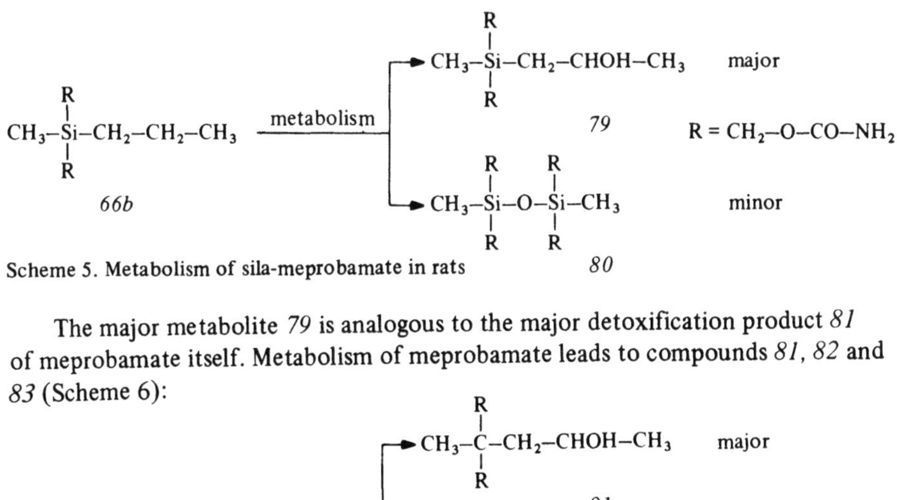

Scheme 5. Metabolism of sila-meprobamate in rats

The major metabolite *79* is analogous to the major detoxification product *81* of meprobamate itself. Metabolism of meprobamate leads to compounds *81, 82* and *83* (Scheme 6):

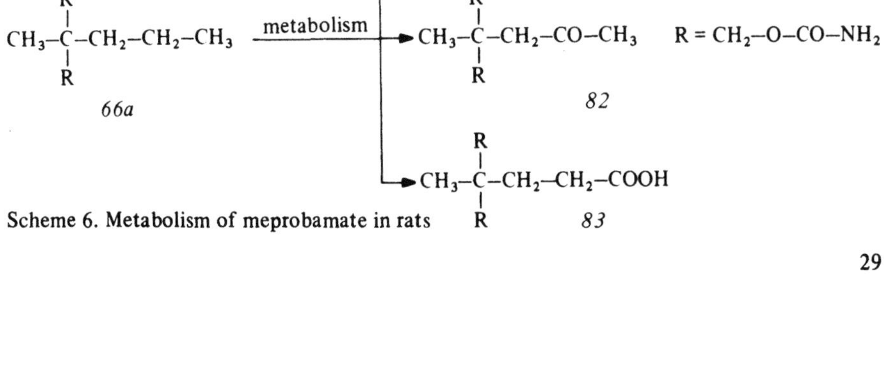

Scheme 6. Metabolism of meprobamate in rats

Although evidence is lacking, Fessenden and Ahlfors supposed that the dealkylated metabolite *80* was actually silanol *84*, which would readily condense to *80* under conditions employed for isolation. Silanol *84* is perhaps a dealkylation product, which is generated by "β-decomposition" after an oxidative process β to silicon:

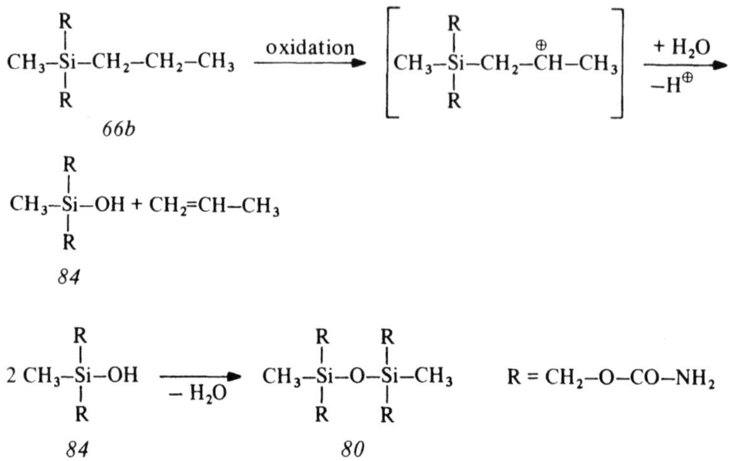

Scheme 7. Rationalization of the formation of metabolite *80*

Investigation of the metabolic fate of compounds *64b* and *71b* led to the following results: After peroral administration of *71b* in rats, about 90% of the ingested silicon could be detected in urine within 4 days. In the ethyl ether extract of the urine, no unchanged *71b* could be found. In the ethyl acetate extract the dealkylation product *85* was detected (29% of ingested sila-carbamate *71b*):

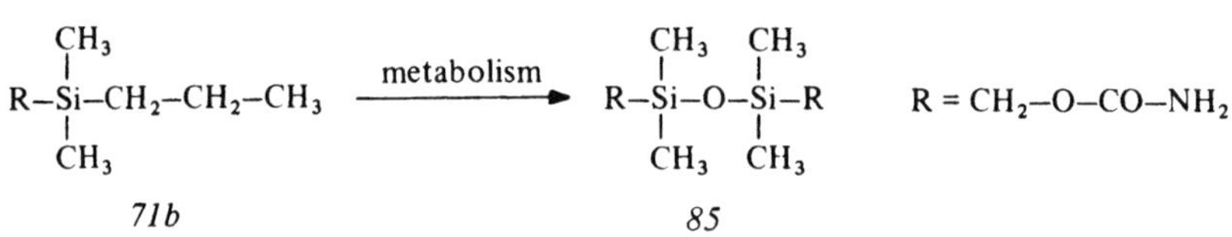

Scheme 8. Metabolism of *71b* in rats

However, no metabolites were observed for compound *64b*. About 70% of the ingested silicon could be detected in urine within 2 days. 53% of the applied sila-carbamate *64b* was isolated from urine as the crystalline unaltered compound.

6.1.2 Metabolisms of Phenyltrimethylsilane, Phenyldimethylsilane, and their Carbon Analogues

Fessenden investigated the metabolic fate[73)] of phenyltrimethylsilane (*86b*) and phenyldimethylsilane (*87b*). The metabolisms of the corresponding carbon analogues t-butylbenzene (*86a*) and isopropylbenzene (*87a*) have been reported earlier by other authors: after being given orally to rabbits, *86a* and *87a* are hydroxylated on the alkyl groups and are eliminated as glucuronide conjugates or as higher oxidized products (cf. Scheme 9). Aromatic ring-hydroxylated metabolites were not reported. However, according to Fessenden preliminary evidence indicates that ring hydroxylation also occurs with *86a*, when given to rats.

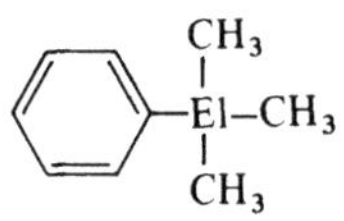

86a: El = C
86b: El = Si

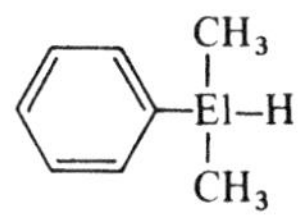

87a: El = C
87b: El = Si

After peroral application to rats silicon compounds *86b* and *87b* were absorbed and subsequently eliminated in the urine as both phenyl- and methyl-hydroxylated metabolites. Metabolic studies of *86b* and *87b* were carried out with the [^{14}C]methyl-silicon label.

After peroral administration of *86b*, 32–40% of radioactivity appeared in the urine within the first 24-hr period. Another 8–20% appeared during the second 24-hr period, and activity decreased to trace amounts after the second day. Three radioactive metabolites, (hydroxymethyl)dimethylphenylsilane (*88*), p-trimethylsilylphenol (*89*), and an unknown conjugate of *90*, could be isolated and characterized (cf. Scheme 9). Metabolism of *87b* leads to silanol *91* (90% in the urine within 36 h).

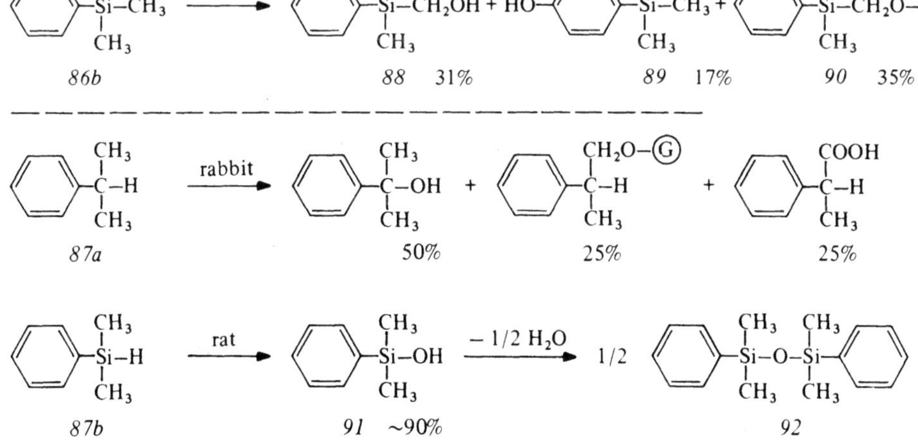

Scheme 9. Metabolisms of phenyltrimethylsilane, phenyldimethylsilane, and their carbon analogues

6.1.3 Silicon Containing Esters with Different Types of Bioactivity

Further examples for relatively small changes in biological activity caused by sila-substitution, were found by Henderson et al.[74], who investigated the effect of acetic acid esters *93a* and *93b* on the isolated guinea pig ileum. Both, the silicon compound *93b* and its carbon analogue *93b* are bioisosters of the neurotransmitter acetylcholine (*94*).

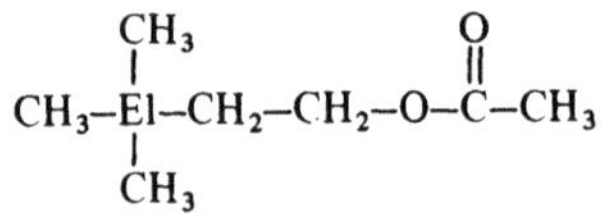

93a: El = C
93b: El = Si

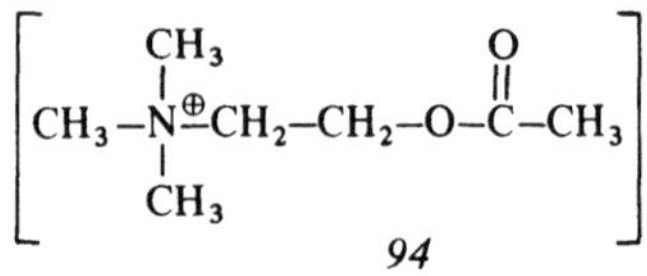

94

The spasmogenic activity of *93a* and *93b* was identified as an indirect cholinergic action. Additionally both esters exhibit a weak anticholinergic and a weak non-competitive papaverine-like action. Great differences between *93a* and *93b* were not observed.

Benzil acid esters *95a* and *95b*, investigated on the isolated heart of the frog, are both relative potent anticholinergic compounds.

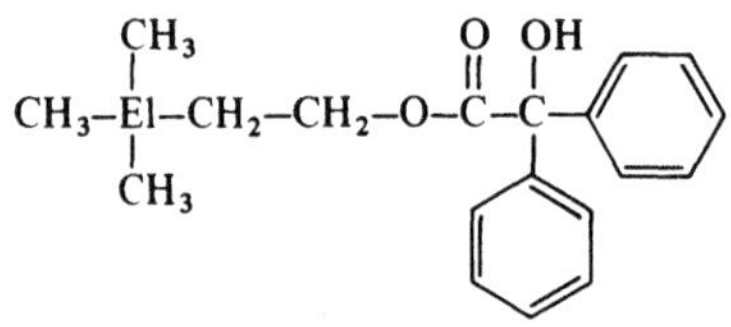

95a: El = C
95b: El = Si

Another study[75], which is concerned with the biological comparison of four pairs of compounds (*96a/96b–99a/99b*), shows also the close relationship between certain silicon compounds and their carbon analogues.

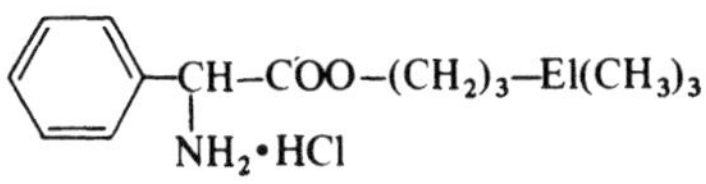

96a: El = C
96b: El = Si

$(CH_3)_3El{-}(CH_2)_3{-}CH(NH_2{\cdot}HCl){-}COO{-}(CH_2)_2{-}CH(CH_3)_2$ *97a*: El = C; *97b*: El = Si

$(CH_3)_3El{-}(CH_2)_2{-}COO{-}(CH_2)_2{-}N(C_2H_5)_2{\cdot}C_6H_8O_7$ *98a*: El = C; *98b*: El = Si

$(CH_3)_3El{-}(CH_2)_3{-}COO{-}(CH_2)_2{-}N(C_2H_5)_2{\cdot}C_6H_8O_7$ *99a*: El = C; *99b*: El = Si

$C_6H_8O_7$: citric acid

Toxicological investigations (white mice, i.p.) of compounds *96a–99a* and *96b–99b* indicated that there is no significant difference in acute toxicity between the silicon compounds and their corresponding carbon analogues (cf. Table 11).

Pharmacological in vitro tests on the isolated guinea pig ileum led to similar results (cf. Table 11). All pairs of compounds tested, with only one exception, are essentially equivalent in their anticholinergic activity: in the case of *96a/96b* the

Table 11. Acute toxicities and spasmolytic activities of *96a/96b–99a/99b*[75)]

Compound	LD_{50}[mg/kg][a]	ED_{50}[Mol/l] acetylcholine	ED_{50}[Mol/l] $BaCl_2$	ED_{50}[Mol/l] histamine
96a	112 (106– 118)	5.21×10^{-5}	2.59×10^{-5}	2.53×10^{-5}
96b	122 (116– 123)	1.44×10^{-5}	3.68×10^{-5}	2.19×10^{-5}
97a	110 (86– 141)	2.84×10^{-5}	2.43×10^{-5}	2.59×10^{-5}
97b	132 (121– 144)	2.64×10^{-5}	3.08×10^{-5}	2.18×10^{-5}
98a	1250 (1164–1343)	9.81×10^{-5}	5.57×10^{-5}	1.36×10^{-5}
98b	1450 (1318–1595)	9.33×10^{-5}	3.62×10^{-5}	1.07×10^{-5}
99a	740 (661– 829)	2.85×10^{-5}	3.25×10^{-5}	1.29×10^{-5}
99b	670 (632– 710)	2.32×10^{-5}	2.48×10^{-5}	1.04×10^{-5}

[a] Confidence limits (95%) in brackets.

silicon compound was found to be four times more active than the corresponding carbon compound. Compared with atropine, all these effects are relatively weak.

Against the spasmogens $BaCl_2$ and histamine, all compounds exhibit relative weak spasmolytic actions. Only in the case of pair *98a/98b* a little, but significant difference in the histaminolytic properties was observed: the silicon compound *98b* acts stronger than the carbon analogue *98a.*

6.1.4 Silicon Containing Amines with Different Types of Bioactivity

Between the sympathomimetic amine *100a* and its silicon analogue *100b* no great differences in biological activity have been observed[76)], although the compounds differ in their acidities (*100a:* $pK_a = 9.73$; *100b:* $pK_a = 10.26$). ECG and EEG of rats,

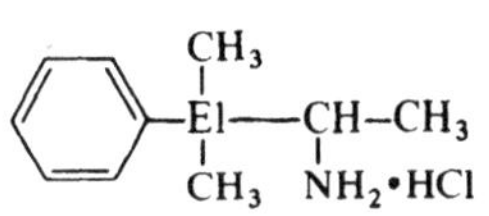

100a: El = C
100b: El = Si

dosed with *100a* and *100b* (sodium pentobarbital sedation), show qualitative similarities. Toxicities (white mice, i.p.) of both compounds are also very similar (*100a:* LD_{50} = 105–113 mg/kg; *100b:* LD_{50} = 102–107 mg/kg). The synthesis of *100b* is described in Scheme 10:

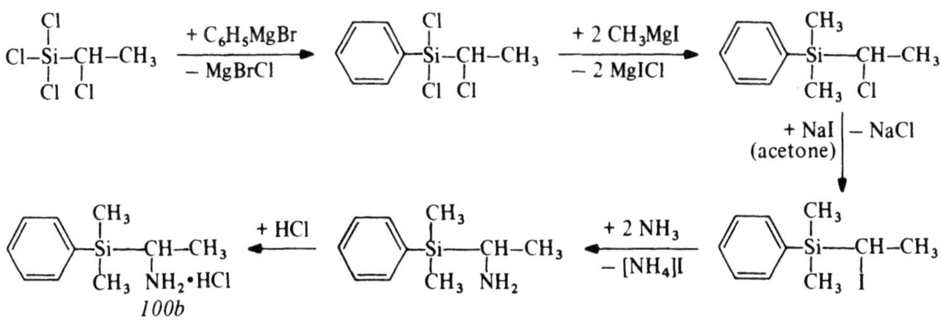

Scheme 10. Synthesis of the sympathomimetic amine *100b*

Further comparative studies[77–79] have been carried out with N-(3-dimethylaminopropyl)-8,8-dimethyl-2-azaspiro[4.5]decane (*101a*) and N-(3-dimethylaminopropyl)-9,9-dimethyl-3-azaspiro[5.5]undecane (*102a*) and their sila-analogues *101b* and *102b*, respectively.

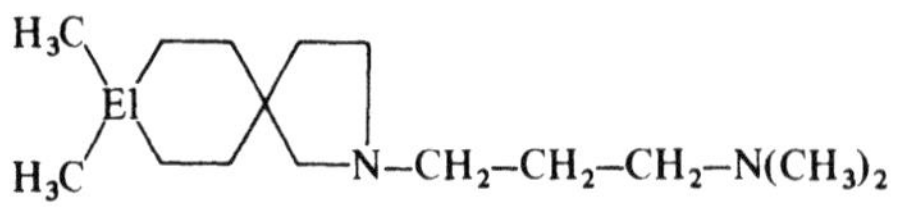

101a: El = C
101b: El = Si

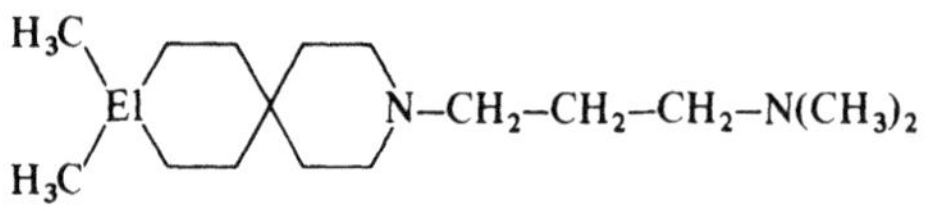

102a: El = C
102b: El = Si

The hydrochlorides of *101a* and *102a* show acute toxicity within the range of 100–125 mg/kg (LD_{50}) in rats. Both compounds inhibit completely cancer cell growth at a concentration of 1×10^{-5} g/ml when tested against human cancer cells grown in tissue culture.

The silicon compounds *101b* and *102b* exhibit a similar biological activity as compared with their carbon analogues: toxicological tests with *101b* and *102b* on rats result in acute LD_{50} range of 100–125 mg/kg. When tested against human cancer cells *101b* inhibits complete cell growth at 1×10^{-5} g/ml, while *102b* inhibits growth at 1×10^{-6} g/ml.

Additionally, the dimethiodide of *101b* was found to lower the blood pressure 40% at a dosage of 10 mg/kg, tested i.v. in dogs under nembutal anaesthesia. Similar tests with the analogous carbon compound are not described.

6.1.5 Silicon Containing Carbamates with Insecticidal Activity

Metcalf and Fukuto have shown[80] that certain silicon compounds interact with enzymes in a manner similar to their carbon analogues. Comparative investigations of *103a* and *103b* on the fly-head cholinesterase resulted in very similar affinities for this enzyme. The silicon compound *103b* acts as inhibitor of fly-head cholinesterase with an affinity for the enzyme 0.57 times that of its carbon analogue *103a*.

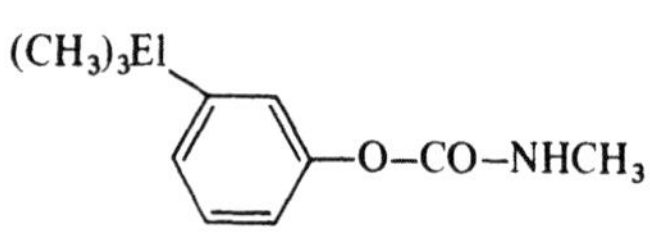

103a: El = C
103b: El = Si

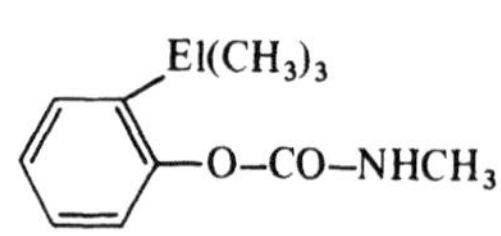

104a: El = C
104b: El = Si

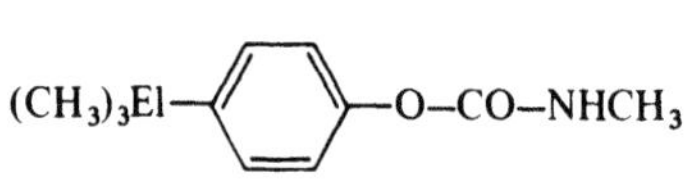

105a: El = C
105b: El = Si

Compound *103b* was found to be more active than the isomeric silicon derivatives *104b* and *105b*. This structure-activity relationship is comparable to that of the corresponding carbon analogues *104a* resp. *105a*.

Comparative investigations of insecticidal properties of the analogous compounds *103a* and *103b* against Culex pipiens quinquefasciatus and Musca domestica also led to similar results.

6.1.6 Silicon Containing Organophosphorus Compounds with Anticholinesterase Activity

The toxicity of organophosphoric esters for insects and mammals is associated with inhibition of cholinesterases. Investigations on the relation between chemical structure of organophosphoric esters and the inactivation of acetylcholinesterase (AChE) have revealed that anticholinesterase activity depends to a large extent on the chemical reactivity of the esters. As a rule, the chemical reactivity of the phosphorus atom is the single most important property which determines the anticholinesterase activity of an organophosphoric ester.

Organophosphates phosphorylate AChE by virtue of an electrophilic attack of the P atom on a serine hydroxyl of AChE, as shown in Eq. (21) for paraoxon (*106*):

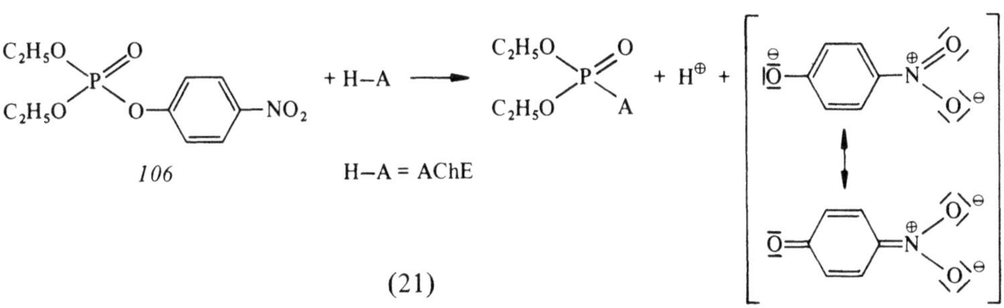

(21)

An electron withdrawing substituent attached to the P atom is an important requirement to give a sufficient electrophilic character. Very potent phosphorylating agents are those which contain leaving groups leading to resonance-stabilized anions. A good example is the p-nitro-phenoxy group in paraoxon [cf. Eq. (21)].

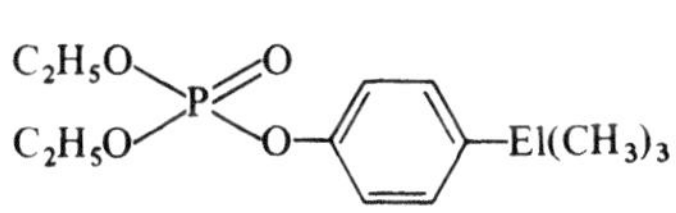

107a: El = C
107b: El = Si

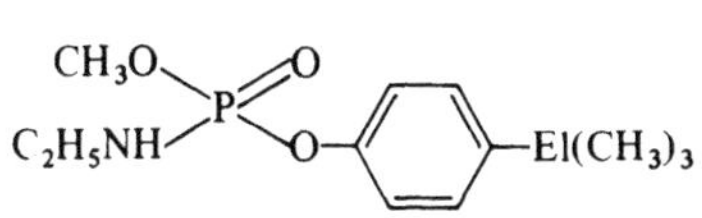

108a: El = C
108b: El = Si

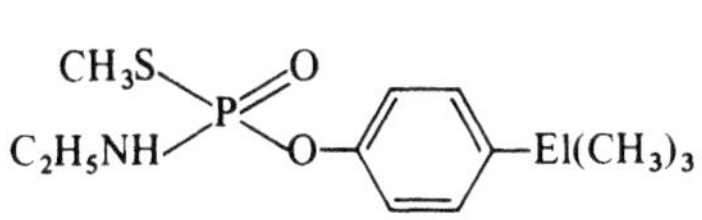

109a: El = C
109b: El = Si

Syntheses (cf. Scheme 11) and comparative biological tests of silicon containing phosphoric esters *107b–109b* and their carbon analogues *107a–109a* were carried out[81]. Because of $(p{\rightarrow}d)_\pi$ interactions between silicon and aryl groups (cf. Chap. 2), the p-trimethylsilylphenoxy group should be a better leaving group for phosphorylation than the p-tert-butylphenoxy group. According to this concept, the silicon compounds *107b–109b* should exhibit a greater anticholinesterase activity than their corresponding carbon analogues *107a–109a*. Preliminary biochemical in vitro tests with *107a* and *107b* seem to confirm this hypothesis: After incubation in human serum for 5 minutes the following ED_{50} values were found (in brackets: 95% confidence limits):

107a: ED_{50} = 1044 (442.6–2466.4) mmol/l human serum
107b: ED_{50} = 158.1 (58.8–424.8) mmol/l human serum
The silicon compound *107b* acts 6.6 times more strongly than its carbon analogue *107a*. Paraoxon was found to be more active than the silicon compound *107b* [*106:* ED_{50} = 20.7 (7.7–55.4) mmol/l human serum]. Further investigations are in progress.

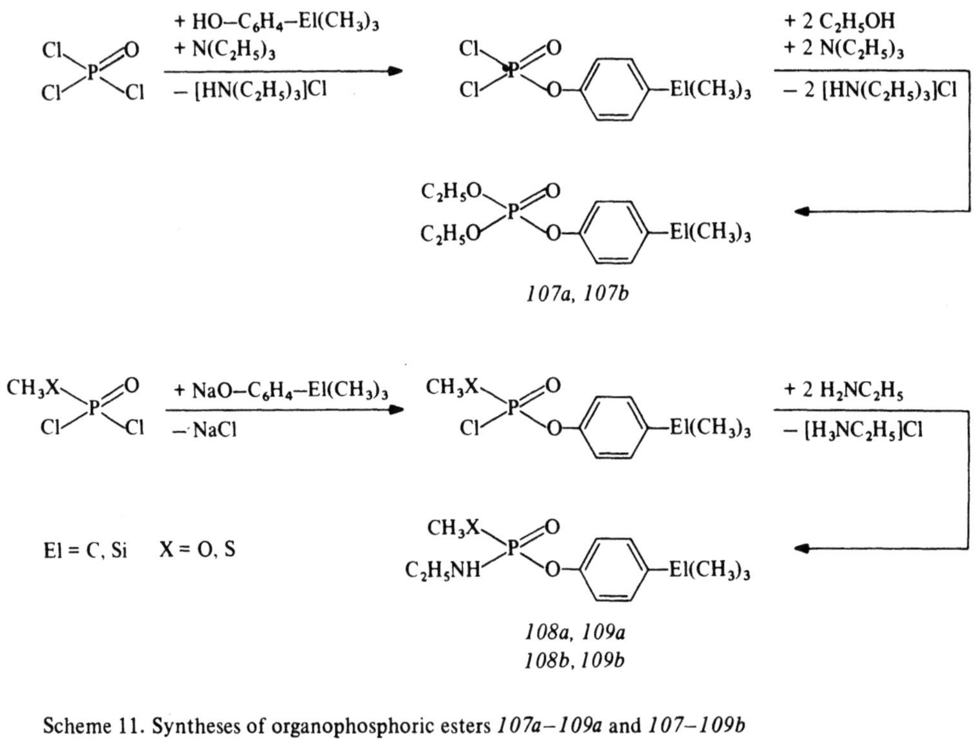

Scheme 11. Syntheses of organophosphoric esters *107a–109a* and *107–109b*

6.1.7 Silicon Containing Derivatives of the Antiseptic Phenoctid and the Ganglionic Stimulant Phenoxyethyl-trimethylammonium

Disila-phenoctid (*110b*), containing two silicon atoms instead of carbon, is a silicon analogue of the antiseptic phenoctid (*110a*). Synthesis of *110b* was achieved accord-

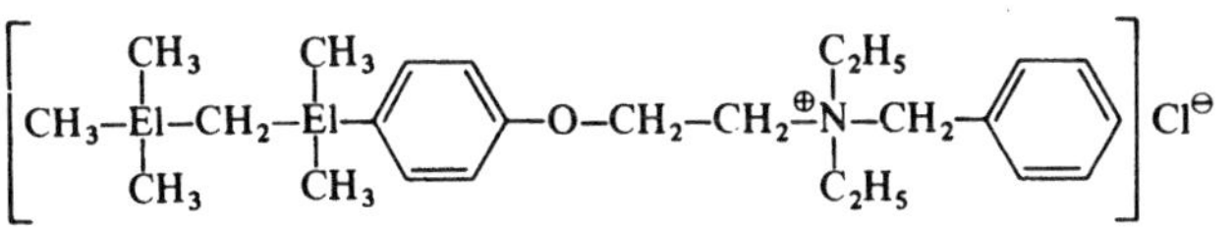

110a: El = C
110b: El = Si

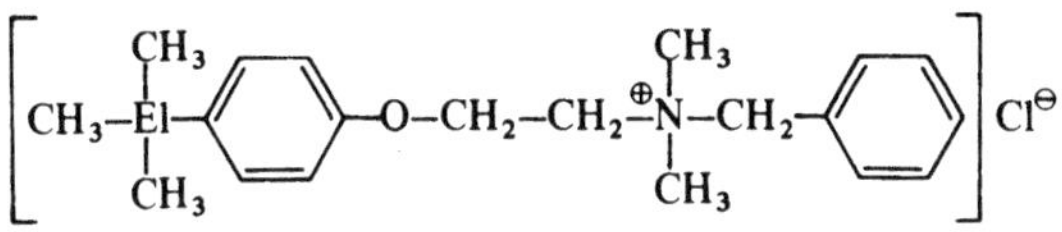

111a: El = C
111b: El = Si

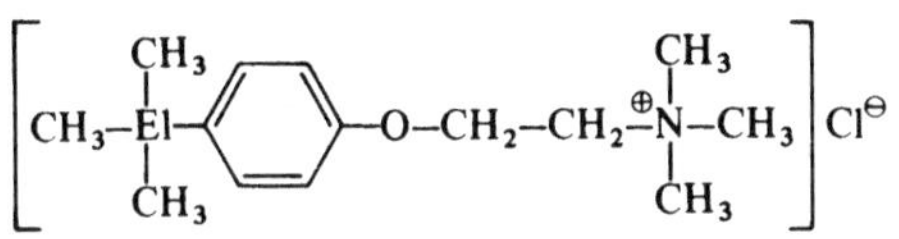

112a: El = C
112b: El = Si

ing to Scheme 12[82]. The silicon containing derivatives *111b* and *112b* were prepared similarly[82].

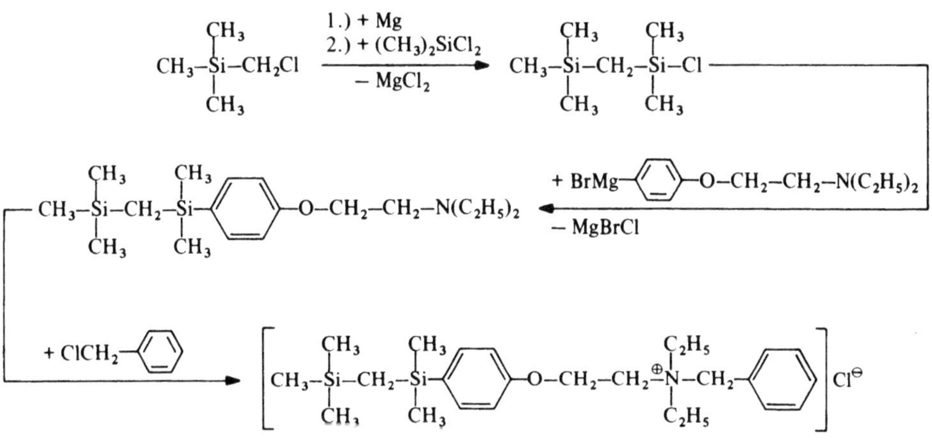

Scheme 12. Synthesis of disila-phenoctid *110b*

Preliminary biological tests[82] have shown that the sila-analogues *110b, 111b* and *112b* exhibit similar antimicrobial activity as compared with that of the corresponding carbon compounds *110a, 111a* and *112a*. Great differences have not been found so far.

Additionally, pair *112a/112b* was tested on the isolated guinea pig ileum. Both compounds, which can be regarded as derivatives of the ganglionic stimulant phenoxyethyl-trimethylammonium (*113*), were found to exhibit similar anticholinergic activity.

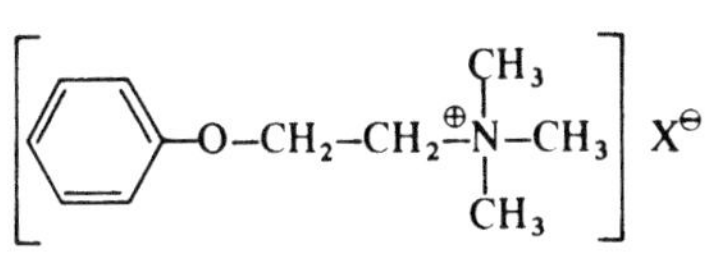

113

6.1.8 (p-Hydroxyphenyl)silanes with Different Types of Bioactivity

(p-Hydroxyphenyl)silanes normally exhibit a greater acidity than analogous carbon compounds, because of $(p \rightarrow d)_\pi$ interactions between aromatic groups and the silicon atom (Chap. 2). This behaviour is possibly the cause of differences in biological activity, if this physico-chemical property influences the pharmacodynamics and/or the specific (or nonspecific) action of the compounds on the receptor (or on cell membranes).

The bactericidal and fungicidal activity of p-trimethylsilylphenol (*114b*) was already reported in 1953[83]. This compound (syntheses[83–85] are described in Scheme 13) is a silicon analogue of the antiseptic p-tert-butylphenol (*114a*):

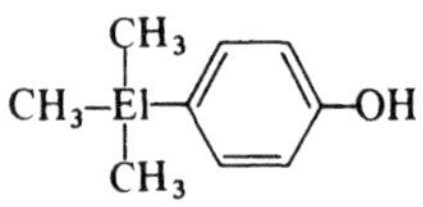

114a: El = C
114b: El = Si

Comparative studies of *114a* and *114b* with respect to their biological properties have not been described in the literature up to now. A derivative of *114b*, bis-(p-

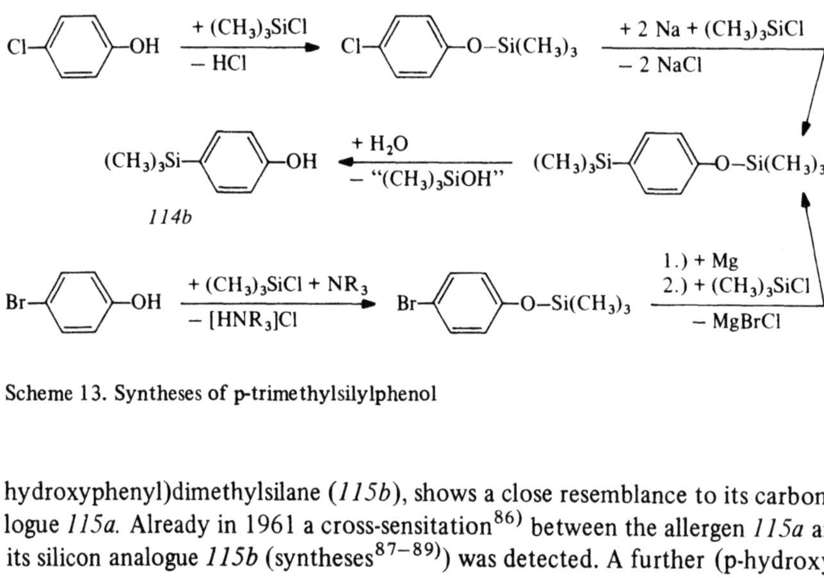

Scheme 13. Syntheses of p-trimethylsilylphenol

hydroxyphenyl)dimethylsilane (*115b*), shows a close resemblance to its carbon analogue *115a*. Already in 1961 a cross-sensitation[86] between the allergen *115a* and its silicon analogue *115b* (syntheses[87–89]) was detected. A further (p-hydroxyphenyl)-

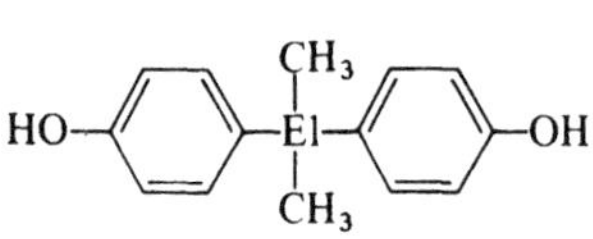

115a: El = C
115b: El = Si

silane which was investigated[90] with respect to its biological activity is sila-rythmol (*116b*). This silicon compound (the synthesis is described in Scheme 14) is an analogue of the antiarrhythmic rythmol (*116a*):

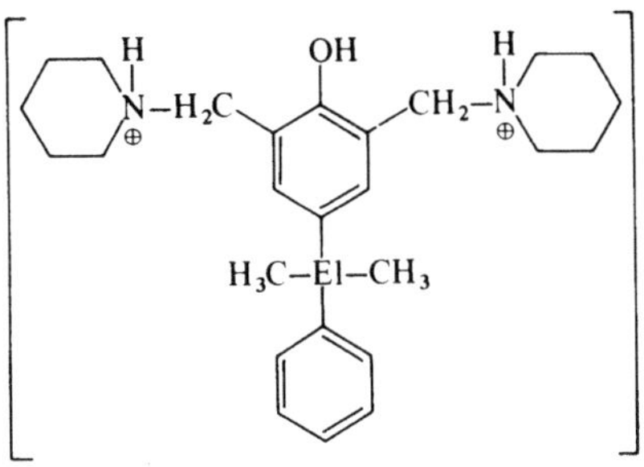

116a: El = C
116b: El = Si

Comparative in vitro studies[90] on the isolated left auricle of guinea pigs have shown that both compounds exhibit a depression of contractility of the heart and a prolongation of the refractory period. However, in comparison to the antiarrhythmic ajmaline, the increase of duration of the refractory period is relatively weak. Striking differences in the pharmacological action of the analogues *116a* and *116b* were not observed.

However, comparative toxicological investigations[90] led to significant differences in their acute toxicities. After i.p. application (white mice), the following LD_{50} values were found:

116a: LD_{50} = 136 mg/kg; *116b:* LD_{50} = 85 mg/kg

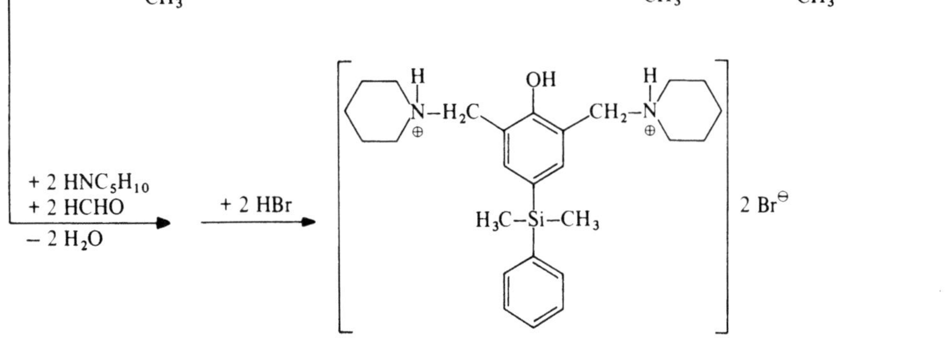

116b

Scheme 14. Synthesis of sila-rythmol

6.1.9 Tricyclic Silicon Containing Compounds with Potential Psychotropic Activity

Compounds of type *117*, containing a tricyclic nucleus with a six-or seven-membered central ring A, a three atom aliphatic side chain (saturated or unsaturated, linear or branched) bound to the central ring A, and a basic functional group (ter-

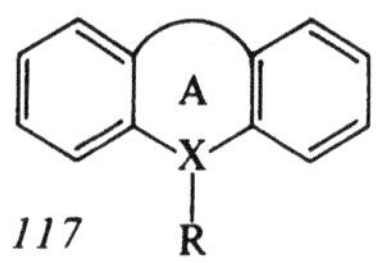

117

>X– = >N–, >C=
R = amino-substituted C_3 side chain
A = six– or seven–membered ring

tiary or secondary amine) bound to the side chain, exhibit psychotropic activity. Compounds of this type of structure are widely used clinically as neuroleptics [e.g. promazine (*118*)] and antidepressants [e.g. imipramine (*119*)].

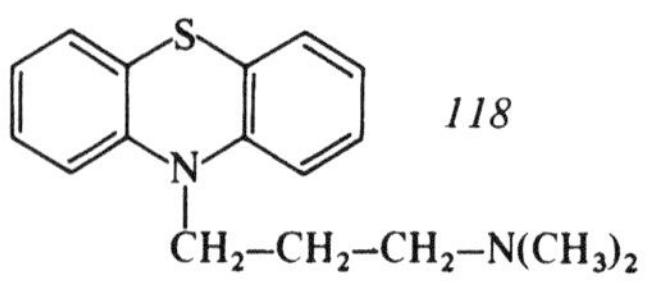

118

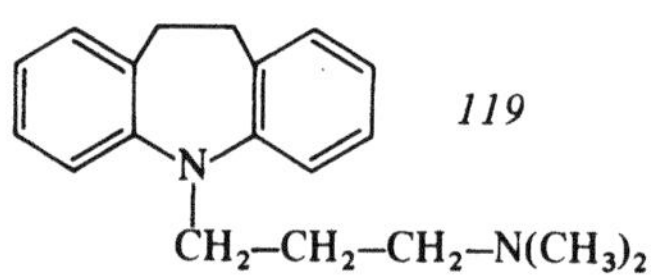

119

According to the theory of Stach and Pöldinger[91], there is a relationship between stereochemical structure and character of the psychotropic action of dibenzo tricyclic drugs: compounds of type *117* with non-planar tricyclic nuclei exhibit predominantly thymoleptic action, whereas compounds with more or less planar frameworks show chiefly neuroleptic activity. Wilhelm[92] also proposes that the main psychotropic activity of tricyclic pharmaceuticals is related to the dihedral angle α (Fig. 2), which is defined as the angle between the planes of the two benzo groups. According to this hypothesis, relatively flat molecules ($\alpha = 145 \pm 10°$) produce a neuroleptic effect, whereas a more marked flexure ($\alpha \sim 125°$) leads chiefly to antidepressant properties.

In addition to the stereochemical structure of the tricyclic skeleton, the conformation of the amino-substituted side chain can influence the psychotropic activity of such compounds. According to the theory of Wilhelm, transmission of the basic psychotropic activity depends largely on the constellation adopted by the side chain. However, the main psychotropic action (neuroleptic or thymoleptic) is chiefly a function of the particular stereochemistry of the tricyclic framework.

Some novel experimental results are in contradiction to the hypothesis of a stereochemical structure-activity relationship. Two research groups (Corey et al., Tacke, Wannagat et al.) are working independently from each other on silicon containing tricyclic compounds with potential psychotropic activity, to bring some light to this dilemma.

Silicon analogues *120b*[93–96] and *121b*[93–95, 97] of the antidepressants dimetacrine (*120a*) and melitracene (*121a*) have been synthesized to study the biological effects generated by sila-substitution in the central ring A (cf. Fig. 2). Sila-substitution

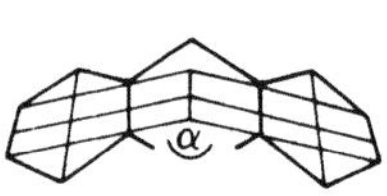

Fig. 2. Steric parameter of tricyclic psychotropic drugs: dihedral angle α; taken from Ref. [92]

in the dimetacrine framework and melitracene skeleton should lead to different stereochemical structures for the sila-analogues *120b* and *121b*, because of the greater Si-C bond length (compared with the C-C bond length) and the ability of the silicon atom to interact with the π-electrons of aryl substituents (Chap. 2). Different constellations of the 3-dimethylaminopropyl groups are also conceivable.

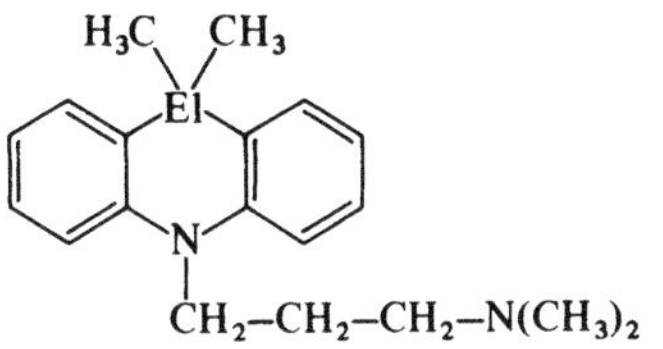

120a: El = C *120b*: El = Si

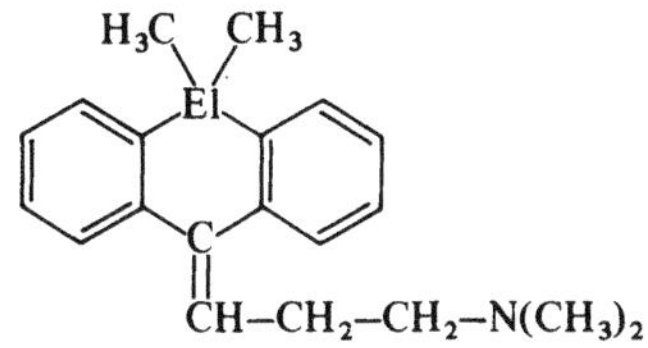

121a: El = C *121b*: El = Si

According to the concept of Stach/Pöldinger[91] and Wilhelm[92], such alterations in stereochemical structure should result in changed psychotropic activities. To check this theory, comparative X-ray diffraction analyses and comparative biological tests are under investigation.

Syntheses of sila-dimetacrine and sila-melitracene are quite different from those of their carbon analogues, as can be seen from Schemes 15 and 16 respectively:

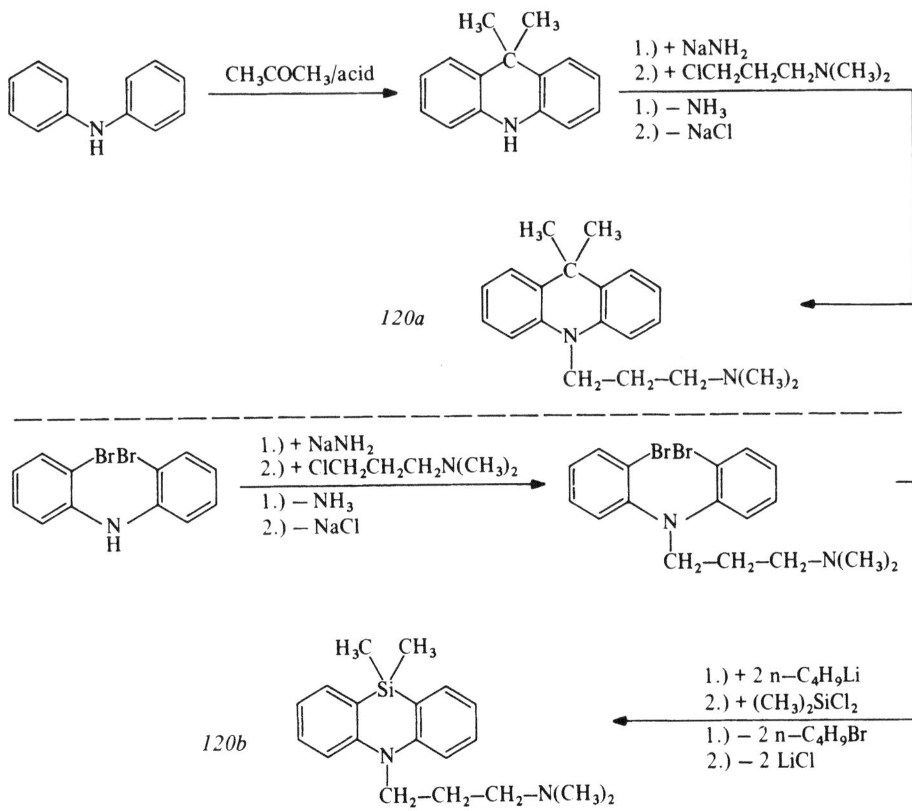

Scheme 15. Syntheses of dimetacrine and sila-dimetacrine

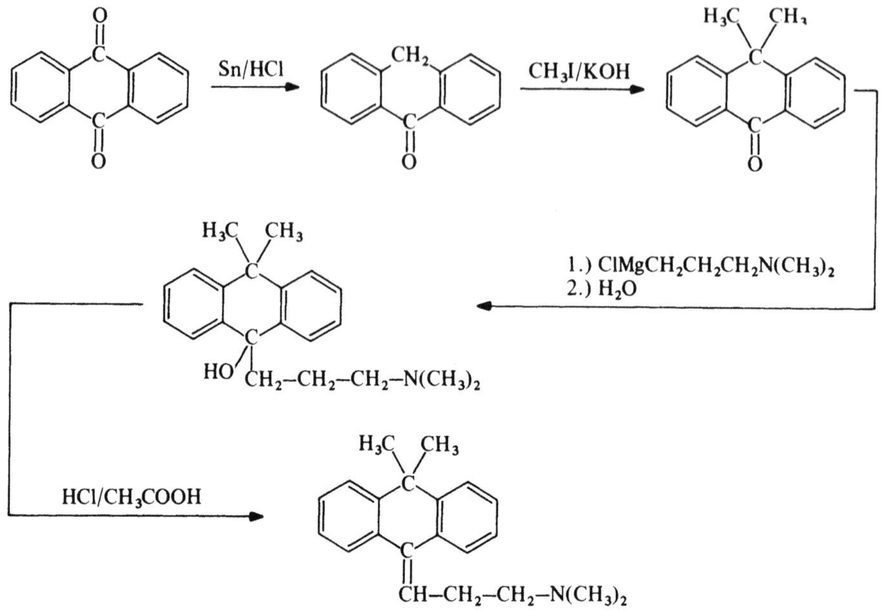

121a

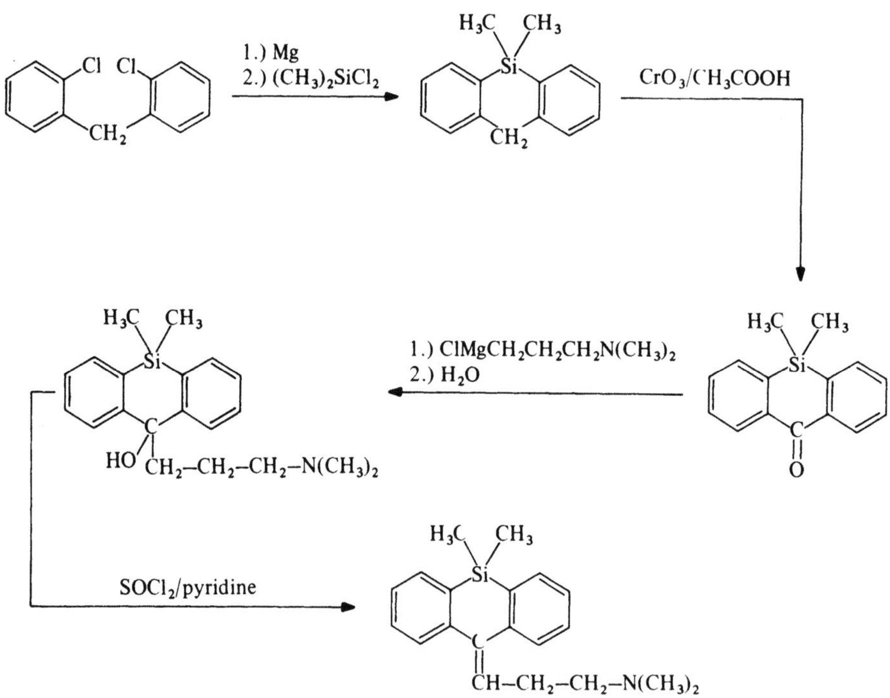

121b

Scheme 16. Syntheses of melitracene and sila-melitracene

Side chain derivatives *122* – *128* of sila-dimetacrine were synthesized[93, 94] analogously to *120b*.

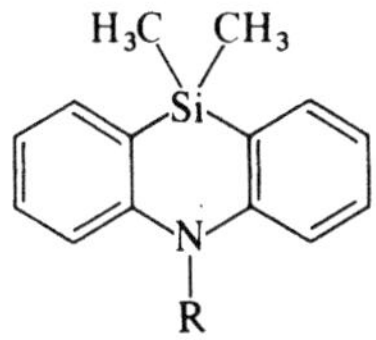

122–128

122: R = CH_2–CH_2–CH_2–N

123: R = CH_2–CH_2–CH_2–$N(C_2H_5)_2$

124: R = CH_2–CH(CH_3)–CH_2–$N(CH_3)_2$

125: R = CH_2– … CH_3

126: CH_2–CH_2–$N(CH_3)_2$

127: CH_2–CH_2–N

128: CH_2–CH_2–N

Additionally, further derivatives of sila-dimetacrine [compounds with other substituents bound to the silicon bridge atom (*129–131*), compounds with a seven-membered central ring containing a Si-Si bridge (*132–134*), compound *135* with a chlorine atom bound to the tricycle] and sila-melitracene (compounds *136* and *137*) were synthesized[93, 94].

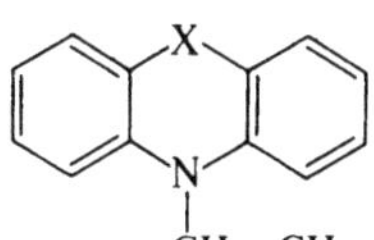

CH_2–CH_2–CH_2–$N(CH_3)_2$

129–131

129: X = $Si(CH_3)C_6H_5$
130: X = $Si(CH_3)H$
131: X = $Si(CH_3)OH$

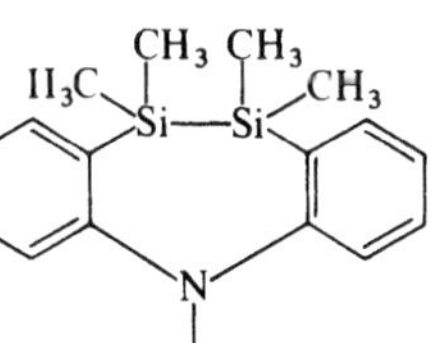

CH_2–CH_2–CH_2–NR_2

132–134

132: NR_2 = $N(CH_3)_2$
133: NR_2 = $N(C_2H_5)_2$
134: NR_2 = N

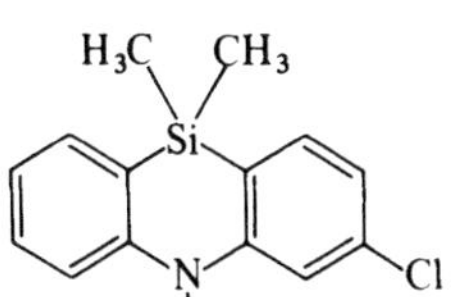

CH_2–CH_2–CH_2–$N(CH_3)_2$

135

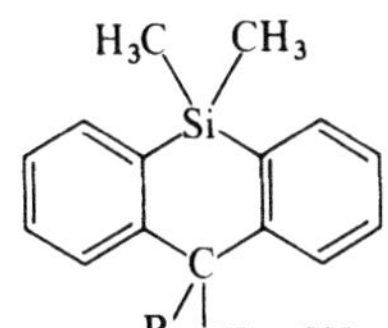

CH_2–CH_2–CH_2–$N(CH_3)_2$

136, 137

136: R = H
137: R = OH

Corey et al. also report compounds (*137–140*)[97] related to sila-melitracene, and dibromo derivative *141*[98], related to sila-dimetacrine.

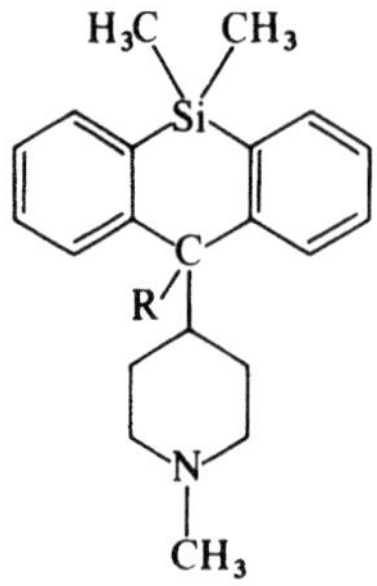

138, 139

138: R = OH
139: R = Cl

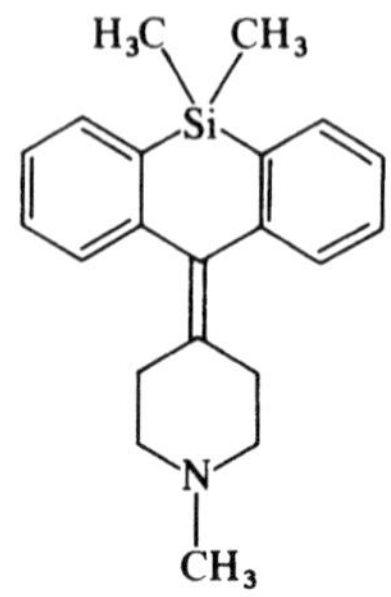

140

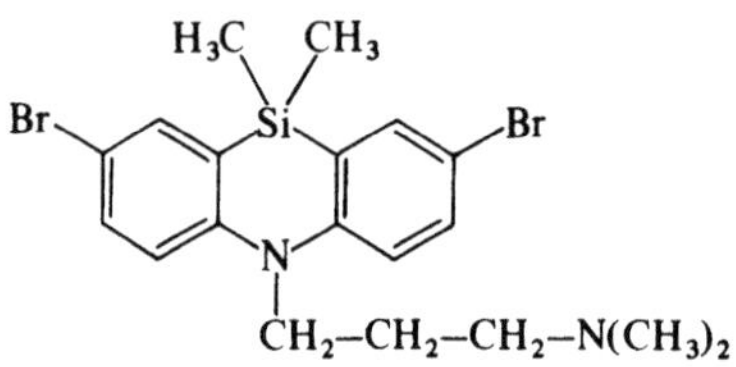

141

Compounds *142–149* were also prepared by Corey et al.[98, 99].

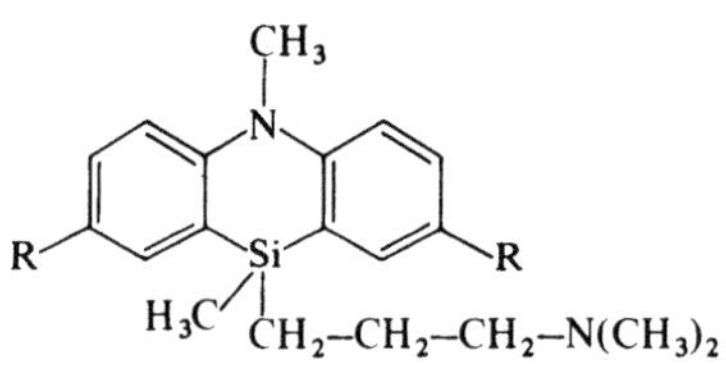

142, 143

142: R = Br
143: R = H

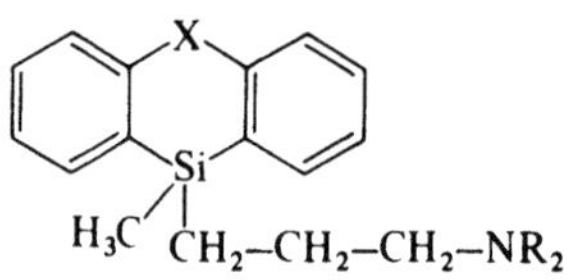

144–147

144: X = CH_2–CH_2, R = CH_3
145: X = CH_2–CH_2, R = C_2H_5
146: X = CH=CH, R = C_2H_5
147: X = CH_2, R = CH_3

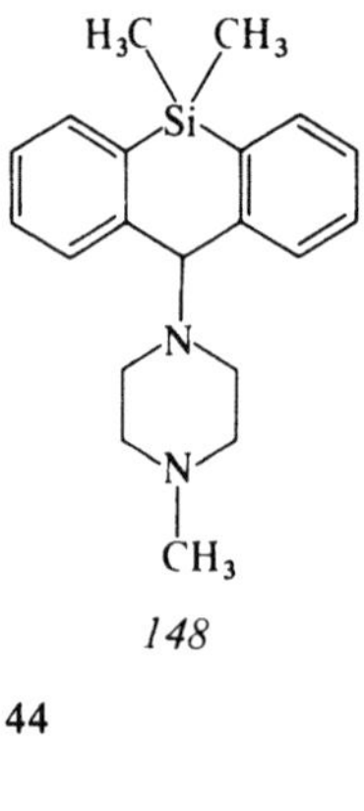

148

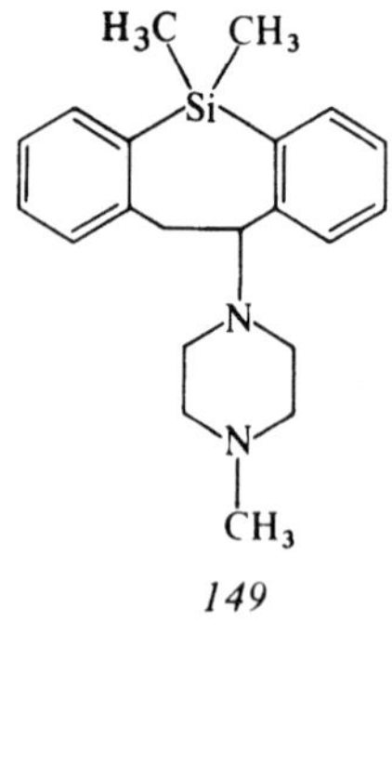

149

Table 12. Comparison of dihedral angle α in the tricyclic compounds *122, 128, 132, 144, 147* and *148*

Compound	Dihedral angle α (°) between benzo groups	Ref.
122	154.1	93)
128	Molecule A 143.9 Molecule B 147.4	93)
132	113.1	93)
144[a]	141.8	100)
147[b]	132.1	101)
148	99	97)

[a] As hydrochloride hydrate.
[b] As hydrochloride.

X-ray diffraction analyses of some tricyclic silicon containing compounds with potential psychotropic activity have been carried out[93, 100, 101]. The experimental dihedral angles α are summarized in Table 12.

The HCl salts of *144, 145, 146, 147* as well as the fumarate salt of *149* have been tested for CNS activity and exhibit similar patterns of activity, however, the potency is less than that of standard pharmacological agents used for comparison. The neutral base *148* was found to be inactive.

Preliminary in vitro studies with dimetacrine (*120a*) and sila-dimetacrine (*120b*) on the isolated guinea pig ileum have shown that both compounds exhibit weak histaminolytic and anticholinergic activities. Comparative studies for psychotropic activity are in progress.

6.2 Comparison of Structures $(\equiv C)_3Si{-}OC\equiv$ and $(\equiv C)_3C{-}OC\equiv$

According to Chap. 2 ether bonds (C–O–C) are very resistant to hydrolytic decomposition, whereas Si–O–C bonds (for example in alkoxysilanes and aroxysilanes) can readily be cleaved by water, particularly in alkaline medium. Replacement of a carbon atom in drugs containing C–O–C units by a silicon atom ($R_3C{-}O{-}CR_3 \rightarrow R_3Si{-}O{-}CR_3$) should lead to bioactive silicon compounds with bioactivities similar to their analogous carbon compounds. But the sila-pharmaca will differ in duration of action because they can be hydrolyzed by the surrounding water of the biological systems. The rate of hydrolysis in such Si–O–C containing compounds will be influenced to an extent depending on the nature of the substituents, bound to the silicon and to the carbon atom.

Sila-pharmaca with a potential site of fracture in their framework could be useful in cases, in which the drug shall act only for a short time (e.g. spasmolytic or analgesic compounds). After effecting in the desired manner the drug will be destroyed by the water of the biological surrounding without help of complex enzymatic systems.

With reference to this point of view a large amount of bioactive silicon compounds with a short duration of action was synthesized and investigated in pharmacological and toxicological effects.

6.2.1 Silicon Containing Derivatives of Antihistamines of the Benzhydrylether Type

Extensive investigations have been carried out in the field of sila-benzhydryl-ethers[102–106, 111–114] [*150b:* sila-mephenhydramine[104, 113]; *151b:* sila-chlor-phenoxamine[105]; *152b:* sila-clofenetamine[105, 114]; *153b:* sila-mebrophen-hydramine[106]; *154b:* sila-meflophenhydramine[111]] and their derivatives[107–110].

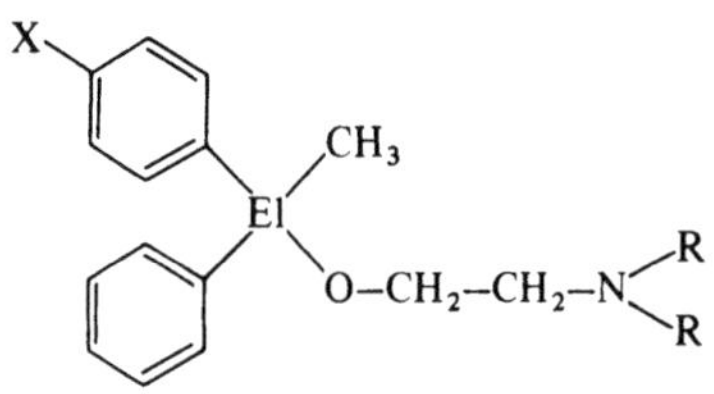

150–154

a: El = C
b: El = Si

150: X = H, R = CH_3
151: X = Cl, R = CH_3
152: X = Cl, R = C_2H_5
153: X = Br, R = CH_3
154: X = F, R = CH_3

Compounds *150b–154b* are silicon analogues (El = Si) of the basic benzhydryl-ether antihistamines *150a–154a* (El = C), which have a great therapeutic and pro-

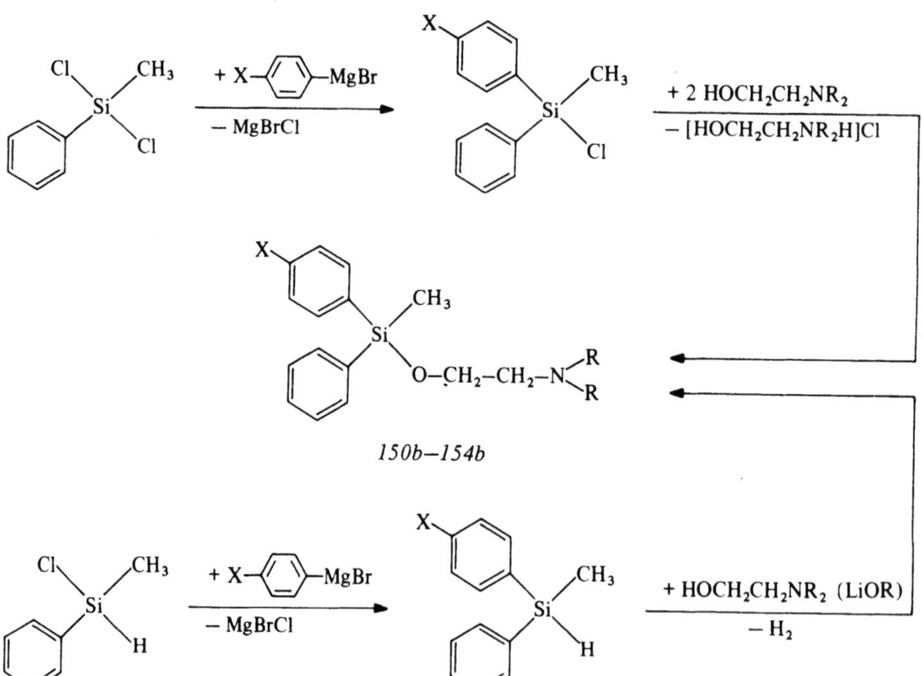

Scheme 17. Syntheses of sila-mephenhydramine, sila-chlorphenoxamine, sila-clofenetamine, sila-mebrophenhydramine and sila-meflophenhydramine

phylactic importance. The carbon compounds act as competitive antagonists of the hormone histamine and the neurotransmitter acetylcholine. They are mainly used for allergies and Parkinson's disease. Many of these compounds also exhibit local anaesthetic, sedative and antiarrhythmic effects.

The syntheses of the sila-pharmaca were achieved through two different routes, as described in Scheme 17.

Many silicon containing derivatives of *150b* (e.g. with $X = CH_3$; NR_2 = morpholino; $Si{-}C_2H_5$ instead of $Si{-}CH_3$; $Si{-}C_6H_{11}$ instead of $Si{-}C_6H_5$; $Si{-}CH_2{-}C$ instead of Si–O–C; Si–NH–C instead of Si–O–C; $Si{-}OCH_2{-}C$ instead of Si–O–C; $Si{-}CH_2O{-}C$ instead of Si–O–C) have been synthesized in a similar manner[104, 106, 108, 109, 111, 114]. The syntheses of the carbon analogues, demonstrated for mephenhydramine (*150a*) and chlorphenoxamine (*151a*), were effected by different means (cf. Scheme 18):

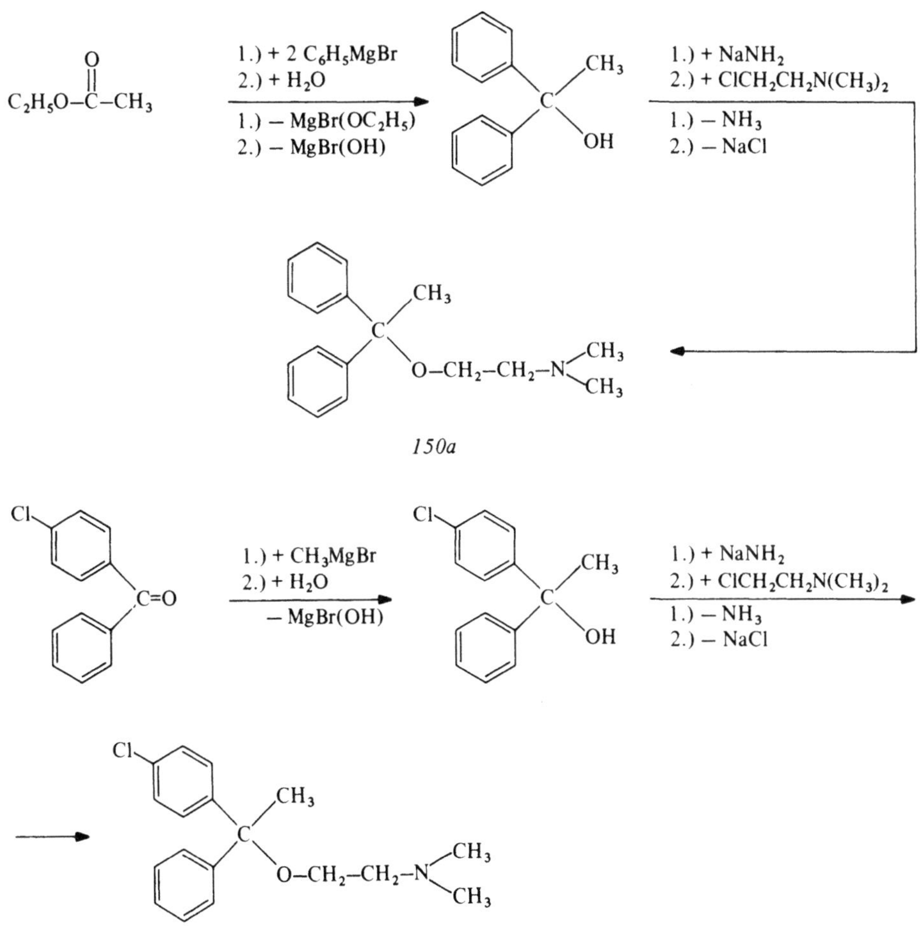

151a

Scheme 18. Syntheses of mephenhydramine and chlorphenoxamine

Table 13. Spasmolytic activities[a] of *150b–153b* and *151a* on the isolated guinea pig ileum

Compound	ED_{50} [mol/l] histamine	ED_{50} [mol/l] carbachol	ED_{50} [mol/l] $BaCl_2$
150b	8.0×10^{-9}	1.6×10^{-7}	3.1×10^{-5}
151b	1.2×10^{-8}	1.3×10^{-6}	2.0×10^{-5}
151a	2.4×10^{-9}	3.7×10^{-7}	1.0×10^{-5}
152b	6.6×10^{-8}	5.2×10^{-7}	1.4×10^{-5}
153b	2.9×10^{-8}	1.6×10^{-6}	–

a Values taken from Ref. [112].

Pharmacological tests[112] (in vitro; guinea pig ileum) of *150b–154b* and some closely related derivatives have shown that the silicon compounds exhibit strong histaminolytic ($ED_{50} \sim 1 \times 10^{-9} - 1 \times 10^{-8}$ mol/l), anticholinergic ($ED_{50} \sim 1 \times 10^{-7} - 1 \times 10^{-6}$ mol/l), and musculotropic effects ($ED_{50} \sim 1 \times 10^{-5}$ mol/l). In most cases the bioactivity of these compounds compares well with that of the corresponding carbon compounds. Some examples are given in Table 13. The sila-analogues exhibit the same structure-activity relationships as the carbon prototypes, but they differ substantially in their duration of action.

Since the Si–O–C bond is sensitive to hydrolysis, compounds *150b–154b* and their derivatives decompose under physiological conditions (Tyrode's solution; pH 7.4; 37 °C) by the action of water, forming the corresponding silanols and amino-alcohols [Eq. (22)]:

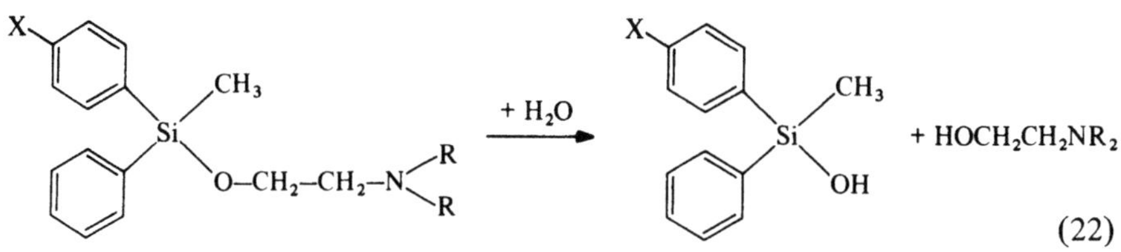

(22)

This results after 15–30 minutes in a quantitative loss of the histaminolytic and parasympatholytic activity, whereas the analogous carbon compounds with stable C–O–C groups decompose after a much longer time by biochemical processes.

The local anaesthetic effects of sila-mephenhydramine and sila-chlorphenoxamine are also of a shorter time as compared with their corresponding carbon compounds[112]. This could also be explained by a hydrolytic inactivation of the sila-pharmaca.

Sila-chlorphenoxamine (LD_{50} = 411.6 mg/kg) exhibits only 25% of acute toxicity (white mice, i.p.) of its carbon analogue (LD_{50} = 108.6 mg/kg)[112]. This observation could easily be explained by a rapid hydrolytic detoxication of the sila-pharmacon.

Sila-mephenhydramine (*150b*), sila-chlorphenoxamine (*151b*) and sila-meflophenhydramine (*154b*) were found to exhibit antiarrhythmic activity on the isolated left auricle of the guinea pig. The silicon compounds lead to an increase of contractility and to a prolongation of the refractory period of the heart. Comparative in vitro studies[112)] with chlorphenoxamine (*151a*) and sila-chlorphenoxamine (*151b*) led to a remarkable difference in activity between carbon compound and sila-analogue: beginning with 10^{-5} mol/l, sila-chlorphenoxamine exhibits an increase of contractility, whereas the analogous carbon compound leads to a depression of contractility (Fig. 3). The behaviour of sila-chlorphenoxamine was found to be an indirect sympathomimetic action. With respect to their influence on the prolongation of the refractory period, both compounds exhibit a similar activity (Fig. 3). In

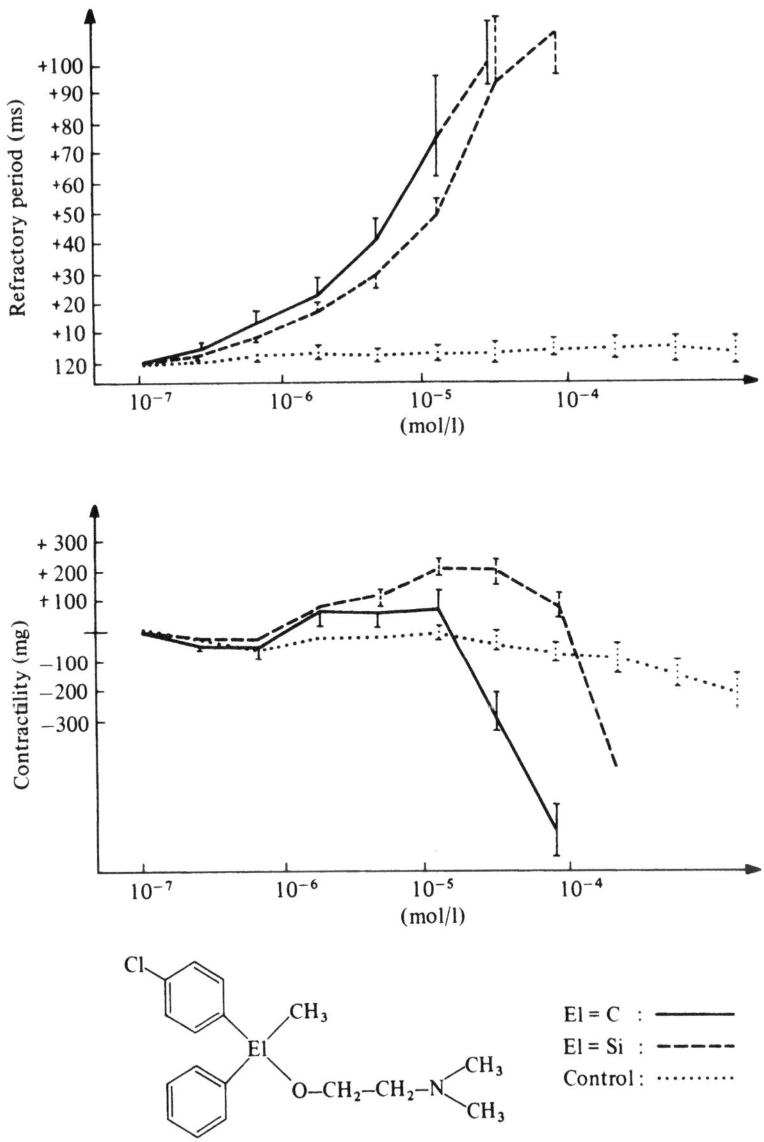

Fig. 3. Dose-activity curves of chlorphenoxamine and sila-chlorphenoxamine

vivo tests with sila-mephenhydramine, sila-clofenetamine, and some of their derivatives have demonstrated the antitremorine activity[113, 114] of these compounds, which is also of a short duration because of hydrolysis.

Substitution of oxygen in the Si–O–C system by the isoelectronic CH_2 group leads to stable compounds with a decreased but prolonged antihistaminic, anticholinergic and antitremorine activity. Replacement of the oxygen atom by a NH group leads to compounds, showing no spasmolytic activity at all because of rapid hydrolysis of the Si–N bond.

Quaternary ammonium salts (type *155*) of the sila-benzhydrylethers are easily prepared by the reaction of the corresponding free bases with CH_3Cl, CH_3Br or CH_3I in CH_3CN[110].

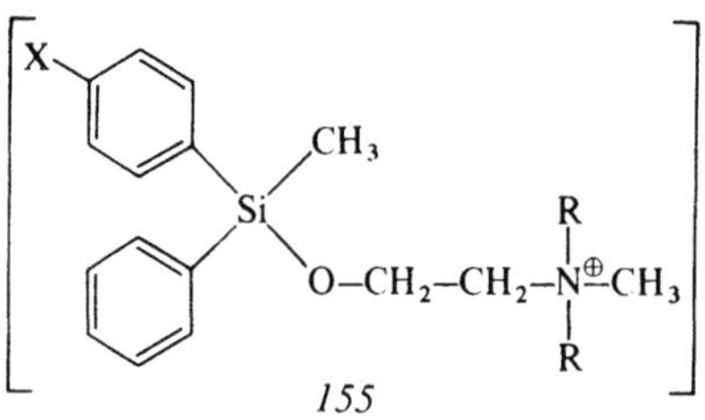

155

$Cl^{\ominus}(Br^{\ominus}, I^{\ominus})$ X = H, Cl, Br
R = CH_3, C_2H_5

Compounds of this type of structure exhibit a stronger anticholinergic and a weaker histaminolytic activity than the corresponding free bases[112]. The same structure-activity relationship is found for analogous carbon systems. Compounds of type *155* exhibit in contrast to their carbon analogue only a short duration of activity (because of hydrolytic inactivation), which is even shorter than that of the basic sila-benzhydrylethers themselves.

Sila-ciclonium bromide (*156b*), which was synthesized[102, 107] according to Scheme 19, is a silicon analogue of the spasmolytic ciclonium bromide (*156a*). It exhibits also a strong spasmolytic activity like the carbon analogue[112]. However, its biological effect is of shorter duration than that of ciclonium bromide itself.

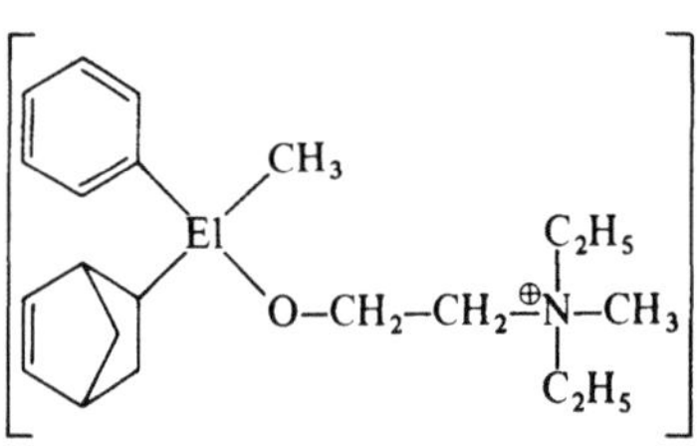

$Br^{\ominus}$ *156a:* El = C
156b: El = Si

Hydrolytic cleavage of the Si–O–C bond leads to an inactivation.

On the isolated guinea-pig ileum both compounds exhibit the same anticholinergic activity (*156a:* $ED_{50} = 6.6 \times 10^{-10}$ mol/l; *156b:* $ED_{50} = 9.1 \times 10^{-10}$ mol/l). However, sila-ciclonium bromide causes a significant larger histaminolytic effect (~4:1) than its carbon analogue (*156a:* $ED_{50} = 1.0 \times 10^{-6}$ mol/l; *156b:* $ED_{50} = 2.2 \times 10^{-7}$ mol/l). The toxicological effects of both compounds after intraperitoneal application (white mice) are nearly the same [*156a:* $LD_{50} = 44.7$ (43.3–46.1) mg/kg; *156b:* $LD_{50} = 40.5$ (39.5–41.9) mg/kg; confidence limits (95%) in brackets].

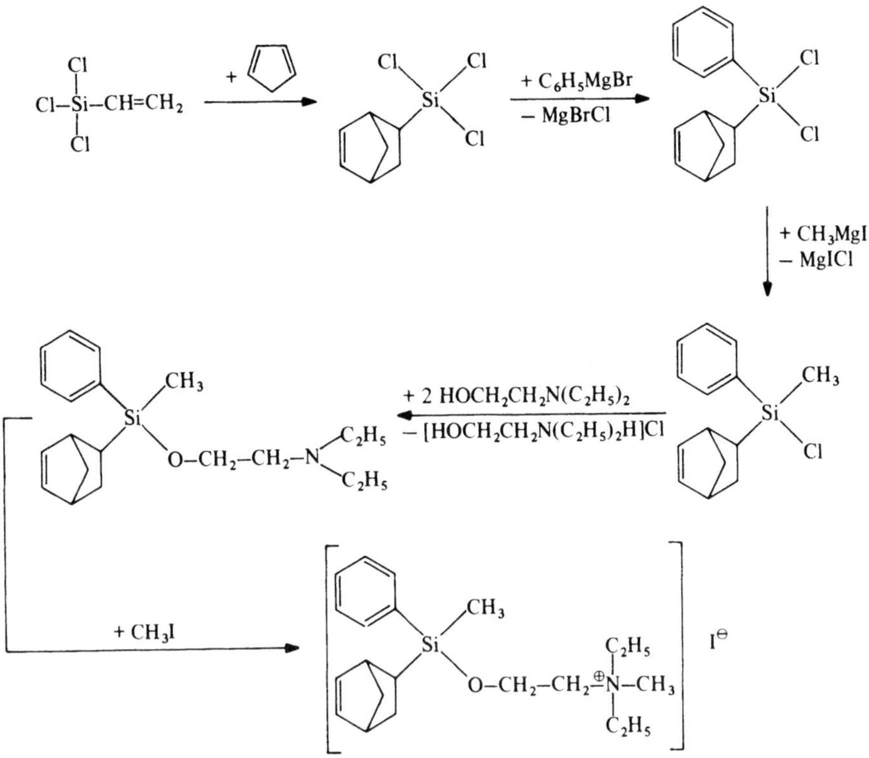

156b

Scheme 19. Synthesis of sila-ciclonium bromide

6.2.2 Silicon Containing Derivatives of the Papaverine-Like Spasmolytics Chlorphencyclane and Bencyclane

In connection with investigations of sila-substituted benzhydrylether antihistamines (cf. Chap. 6.2.1), the syntheses of sila-chlorphencyclane (*157b*), sila-bencyclane (*158b*) and some of their derivatives were carried out[115–117].

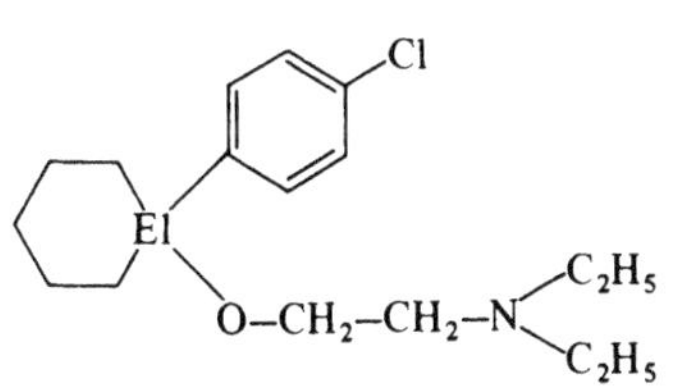

157a: El = C
157b: El = Si

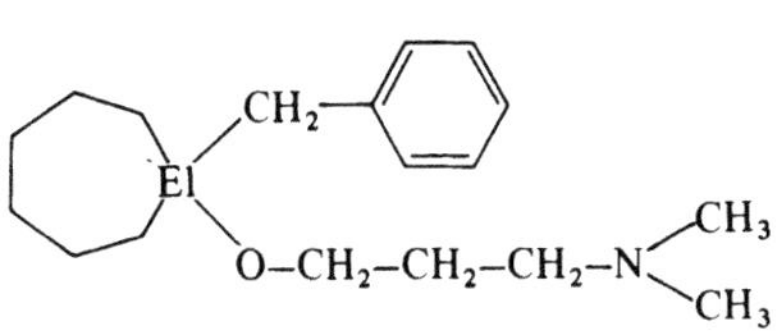

158a: El = C
158b: El = Si

The synthetic routes for sila-chlorphencyclane (*157b*) and its carbon analogue (*157a*) (therapeutic use: central stimulant) are described in Scheme 20:

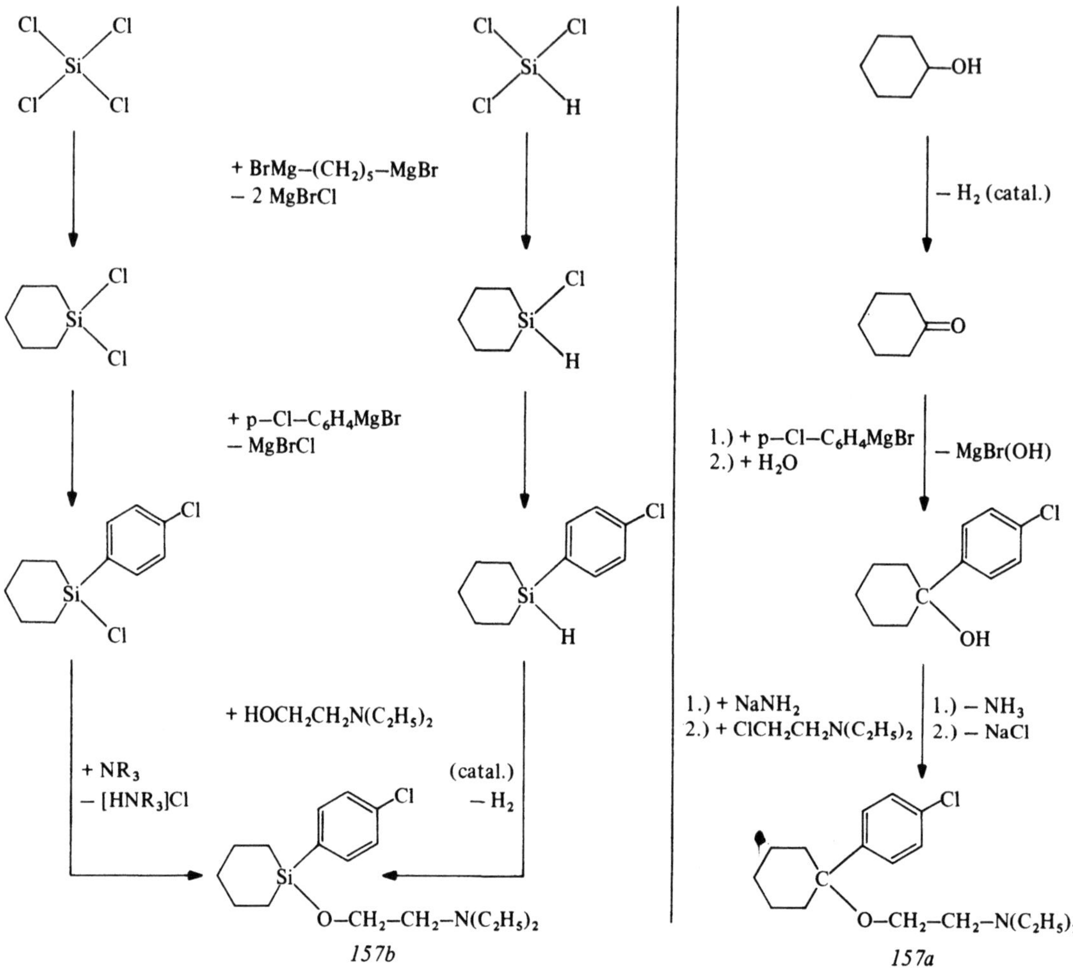

Scheme 20. Syntheses of chlorphencyclane and sila-chlorphencyclane

Some derivatives of sila-chlorphencyclane (replacement of the p-Cl–C_6H_4 group by p-F–C_6H_4, p-CH_3–C_6H_4 or naphtyl; replacement of the $N(C_2H_5)_2$ group by $N(CH_3)_2$ or morpholino; replacement of the sila-cyclohexane ring by a sila-cyclopentane ring; elongation of the OCH_2CH_2 chain to $OCH_2CH_2CH_2$; substitution of the oxygen atom by the isoelectronic CH_2 or NH group or by a sulfur atom; quaternary ammonium salts) have also been synthesized in a similar manner[115–117].

In vitro tests (isolated guinea pig ileum) of the pair *157a*/*157b* and some of their derivatives (*159–167*) were carried out with the spasmogens carbachol, histamine and $BaCl_2$ (Table 14)[118]. Most of the compounds exhibit a spasmolytic activity against $BaCl_2$, which is nearly in all cases of a papaverine-like nature. The histaminolytic and anticholinergic properties of the pair *157a*/*157b* are weak, as compared with the standards diphenhydramine and atropine sulfate.

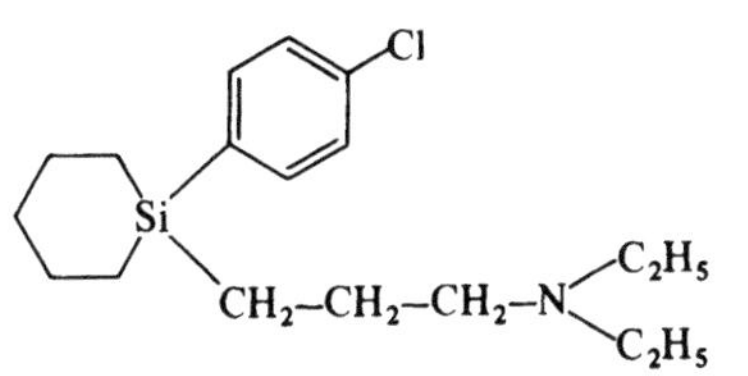

159

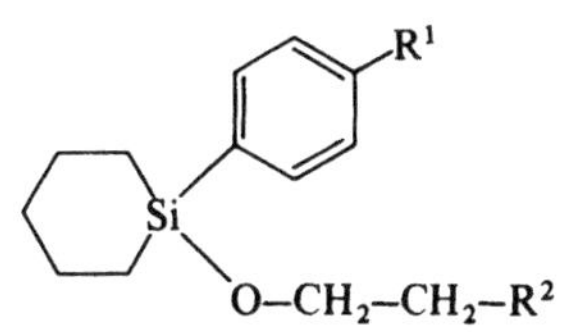

160–167

No.	R^1	R^2	No.	R^1	R^2
160	H	$N(C_2H_5)_2$	*164*	Cl	Morpholino
161	F	$N(C_2H_5)_2$	*165*	Cl	$CH_2N(CH_3)_2$
162	CH_3	$N(C_2H_5)_2$	*166*	Cl	$\overset{\oplus}{N}(C_2H_5)_2CH_3]I^{\ominus}$
163	Cl	$N(CH_3)_2$	*167*	Cl	$\overset{\oplus}{N}(CH_3)_3]I^{\ominus}$

Table 14. Spasmolytic activities[a] of *157a*, *157b* and *159–167*

Compound	ED_{50} [μ mol/l][b] carbachol	ED_{50} [μ mol/l][b] histamine	ED_{50} [μ mol/l][b] $BaCl_2$
157a	0.95 (0.32 – 2.8)	1.9 (0.77– 4.5)	11 (3.2 – 37)
157b	0.84 (0.29 – 2.4)	1.7 (0.60– 4.9)	5.6 (1.8 – 18)
159	0.62 (0.16 – 2.5)	0.75 (0.22– 2.5)	2.0 (0.55– 6.7)
160	0.40 (0.13 – 1.3)	3.8 (1.5 –10)	9.4 (3.2 – 28)
161	0.91 (0.32 – 2.6)	1.3 (0.49– 3.7)	7.0 (2.3 – 22)
162	1.4 (0.45 – 4.1)	2.9 (1.0 – 8.2)	–
163	0.90 (0.32 – 2.6)	1.4 (0.46– 4.5)	5.7 (1.7 – 19)
164	7.6 (2.7 –21)	16 (5.3 –46)	43 (12 –149)
165	0.34 (0.11 – 1.0)	2.1 (0.70– 6.0)	4.4 (1.5 – 13)
166	0.10 (0.035– 0.30)	3.4 (1.3 – 9.0)	14 (3.9 – 50)
167	0.42 (0.17 – 0.99)	8.6 (3.0 –24)	27 (7.1 –104)
Standards			
Atropine sulfate	0.0027 (0.0010–0.0070)	–	–
Diphen-hydramine	–	0.039 (0.016–0.10)	–
Papaverine	–	–	14 (2.9–68)

[a] Values taken from Ref. [118].
[b] Confidence limits (90%) in brackets.

Chlorphencyclane and sila-chlorphencyclane show no significant differences in their spasmolytic activity against the spasmogens carbachol, histamine and $BaCl_2$. Both compounds act more strongly than the standard papaverine. The isoelectronic derivative *159* of sila-chlorphencyclane exhibits even a spasmolytic activity, which is 7 times stronger than that of papaverine. The duration of the spasmolytic action of sila-chlorphencyclane and its Si–O–C containing derivatives is shorter than that of chlorphencyclane, because of hydrolytic cleavage of the Si–O–C bond [Eq. (23)]:

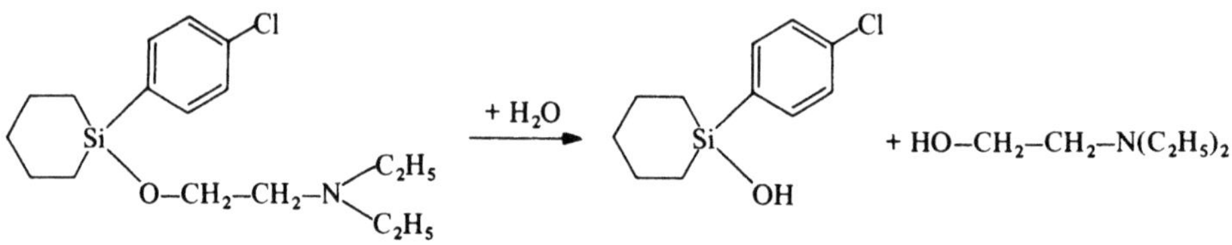

157b *168* (23)

After about 3 hours (measured under physiological conditions in Tyrode's solution at 37 °C and pH 7.4) there is only a slight activity, which corresponds approximately to the silanol *168*. This fragment exhibits also a weak non-specific activity[118)]

In vitro tests at the isolated left auricle of guinea pigs have shown that *157b* and some of its derivatives prolong the functional refractory period, which can be compared approximately with that of the drug quinidine. Between the analogues *157a* and *157b* no significant differences were observed at all with this pharmacological model[118)].

Toxicological investigations[118)] (white mice, i.p.) resulted in a significant difference in LD_{50} values for chlorphencyclane [LD_{50} = 142.1 mg/kg (459 μmol/kg)] and sila-chlorphencyclane [LD_{50} = 185.8 mg/kg (570 μmol/kg)]. However, the hydrochlorides of *157a* [LD_{50} = 102.6 mg/kg (296 μmol/kg)] and *157b* [LD_{50} = = 109.3 mg/kg (302 μmol/kg)] exhibit the same toxic effects. Further toxicological data for *159*, *161*, *164* and *166* are given in Table 15.

Table 15. Toxicological effects[a] of *157a*/*157b* and *159, 161, 164* and *166*

Compound	LD_{50} [mg/kg][b]
157a	142.1 (130.2–155.4)
157b	185.8 (176.0–195.9)
157a as HCl salt	102.6 (96.6–116.7)
157b as HCl salt	109.3 (101.7–117.4)
159 as HCl salt	92.0 (81.7–103.5)
161	215 (197 –235)
164	688 (558 –847)
166	56.7 (51.1– 62.8)

[a] Values taken from Ref. [118)].
[b] Confidence limits (95%) in brackets.

6.3 Comparison of Structures $R_3Si–H$ and $R_3C–H$

In nearly all cases carbon atoms of organic drugs have at least one C–H bond. So it seems not to be very useful to substitute such carbon atoms for silicon atoms, because of the high hydrolytic sensitivity of Si–H bonds (Chap. 2). This fact reduces the possibilities of expanding the field of sila-pharmaca.

6.3.1 Silicon Containing Derivatives of the Antihistamines Diphenhydramine and Fenpiprane

Syntheses and biological investigations of Si–H containing systems were carried out with sila-diphenhydramine (*169b*), sila-fenpiprane (*170b*) and some of its derivatives (*171, 172*)[119–121]. The syntheses[119] of the antihistamine fenpiprane (*170a*) and its sila-analogue *170b* are described in Scheme 21.

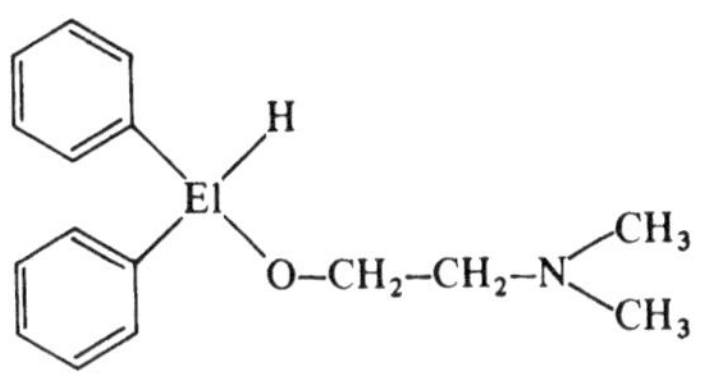

169a: El = C
169b: El = Si

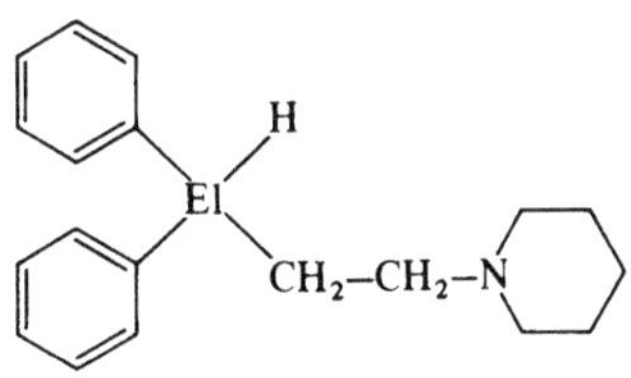

170a: El = C
170b: El = Si

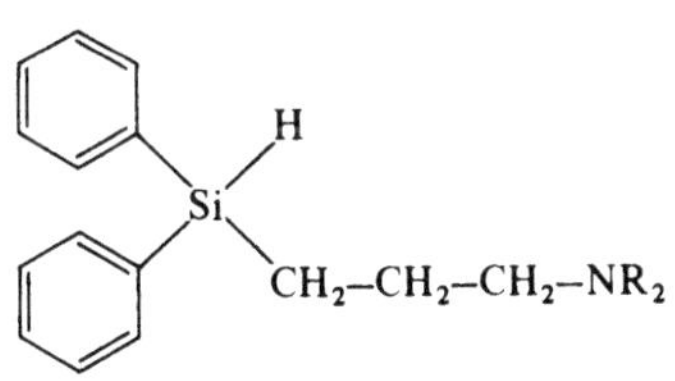

171: $NR_2 = N(CH_3)_2$

172: $NR_2 =$ 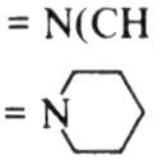

The antihistamine diphenhydramine (*169a*) exhibits strong histaminolytic effects on the isolated guinea pig ileum, whereas its sila-analogue *169b* hydrolyses [Eq. (24)] so fast that characteristic effects could not be detected. This can be explained easily by the presence of the sensitive Si–H and Si–OC bond.

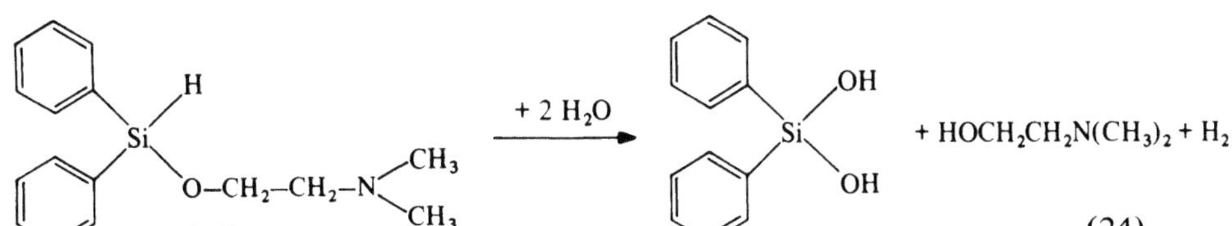

(24)

However, sila-fenpiprane and compounds *171* and *172* are sufficiently stable under physiological conditions that their histaminolytic and anticholinergic effects can be measured. Preliminary in vitro tests[120] with *172* on the isolated guinea pig ileum have shown that the bioactivity of this compound is changed with time after adding to an organ bath: the histaminolytic component becomes weaker, whereas the anticholinergic effect increases during a period of about 2 hours (measured in Tyrode's solution at pH 7.4 and 37 °C). This behaviour can be explained by a hydrolytic transformation (Eq. 25) of the Si–H containing compound to the corresponding silanol *173b* (= sila-difenidol, compare Chap. 6.4.1), which has a stronger anticholinergic and a weaker histaminolytic activity than that of its Si–H precursor.

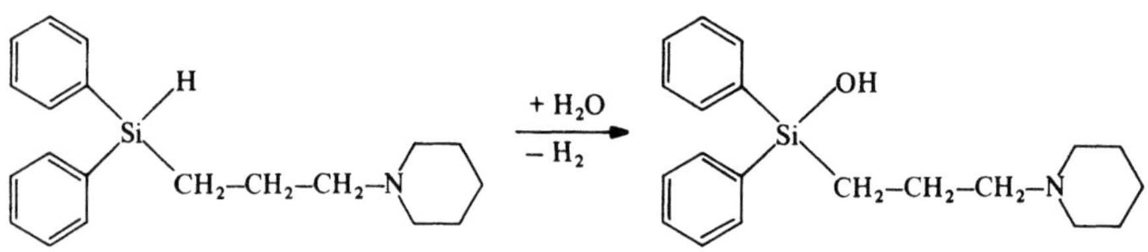

172 *173b* (25)

Comparative in vitro tests[120] on the isolated guinea pig ileum with fenpiprane and sila-fenpiprane lead also to interesting results. Both compounds exhibit spasmolytic activity against the spasmogens histamine, carbachol and $BaCl_2$ (compare Table 16). Remarkable is the fact that sila-fenpiprane shows a significant stronger histaminolytic (~1:36) and a stronger anticholinergic (~1:9) action than fenpiprane itself. During a time of about 15 minutes the histaminolytic activity of sila-fenpiprane decreases, whereas the anticholinergic component increases. This behaviour is also explainable (in analogy to *172*) by a hydrolytic cleavage [Eq. (26)] of the Si–H bond, leading to the potent anticholinergic sila-pridinol (*174b*) (compare Chap. 6.4.1).

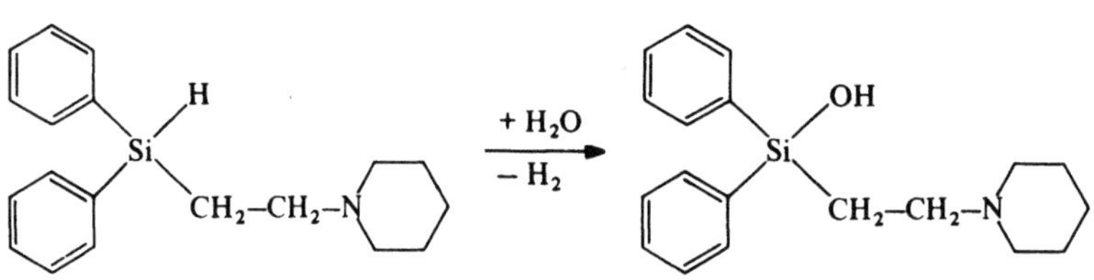

170b *174b* (26)

Toxicological tests[121] (white mice, i.p.) resulted also in significant differences. The silicon compound was found to be more toxic than its carbon analogue. It is not clear so far, whether this difference is connected with the hydrolysis or not.

Summarizing, one can say that Si–H containing compounds can be stable under physiological conditions for a certain time. As a principle, these compounds can exhibit a specific biological activity. Furthermore, Si–H containing compounds could be useful as precursors of potent bioactive silanols. Such precursors, which are more lipophilic than the corresponding silanols, could lead to changed biological properties (as compared with the silanols) with respect to resorption, distribution, and

Table 16. Acute toxicities[121] and spasmolytic activities[120] of *170a* and *170b*[a]

	170a	170b
LD_{50}[mg/kg]	82.0 (73.3–91.7)	62.0 (52.6–73.1)
ED_{50}[mol/l], histamine	1.2×10^{-7} (0.4×10^{-7}– 3.6×10^{-7})	3.3×10^{-9} (0.9×10^{-9}–12.5×10^{-9})
ED_{50}[mol/l], carbachol	2.0×10^{-8} (0.6×10^{-8}– 6.5×10^{-8})	2.3×10^{-9} (0.5×10^{-9}–11.0×10^{-9})
ED_{50}[mol/l] $BaCl_2$	7.9×10^{-6} (2.7×10^{-6}–22.8×10^{-6})	1.9×10^{-5} (0.2×10^{-5}–14.5×10^{-5})

[a] Confidence limits (95%) in brackets.

activity at the receptor, before the stable silanols are formed by hydrolytic cleavage of the Si–H bond. This principle is the opposite of the hydrolytic detoxication of the Si–O–C containing sila-pharmaca, which are described in Chap. 6.2.

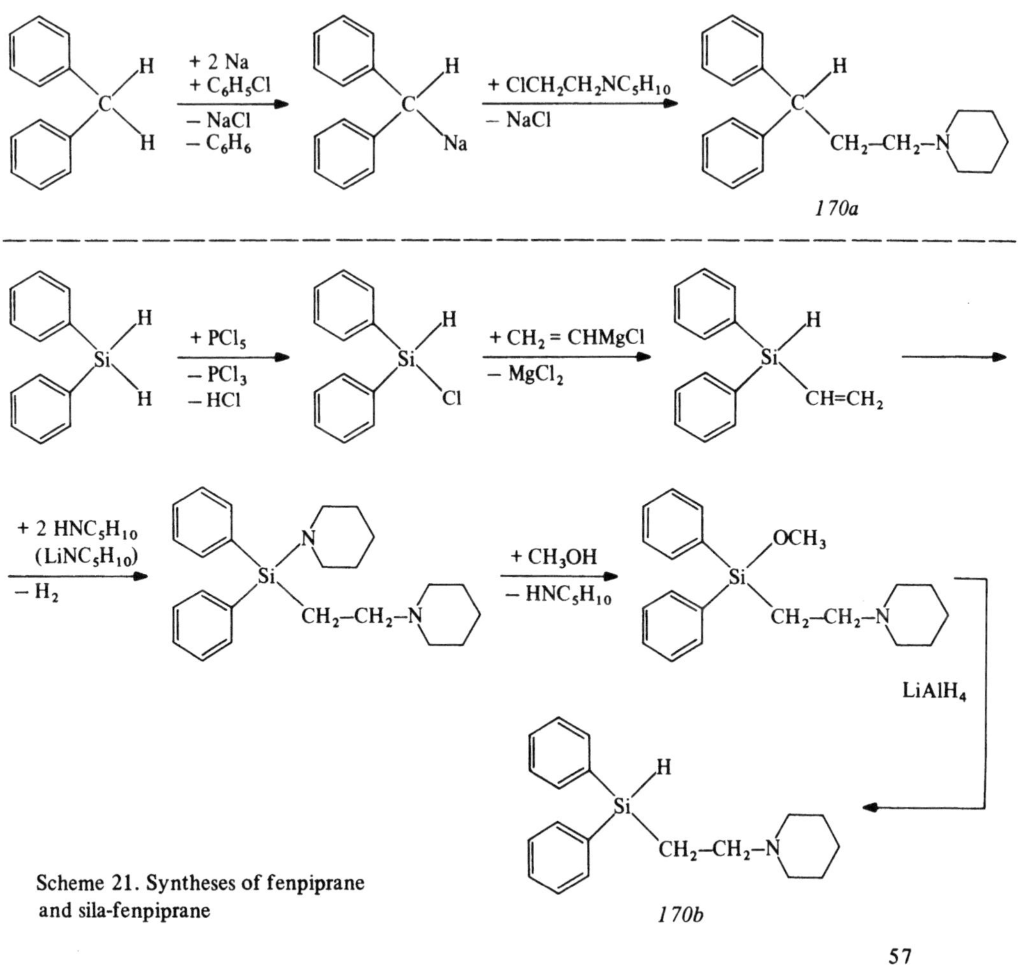

Scheme 21. Syntheses of fenpiprane and sila-fenpiprane

6.4 Comparison of Structures $(\equiv C)_3Si-OH$ and $(\equiv C)_3C-OH$

6.4.1 Silicon Containing Derivatives of Aminosubstituted Tertiary Alcohols with Spasmolytic Activity

The amino-substituted tertiary alcohols *173a–178a* show a therapeutic activity in a widely divergent series of pathologic states. They have a similar structure to that of the benzhydrylether antihistamines (Chap. 6.2.1), but tend to lack strong antihistaminic activity, while they retain some of the sedative and antiparkinson activities. In vitro tests (guinea pig ileum) show a stronger anticholinergic and a weaker histaminolytic action as compared to the benzhydrylether antihistamines. The pro-

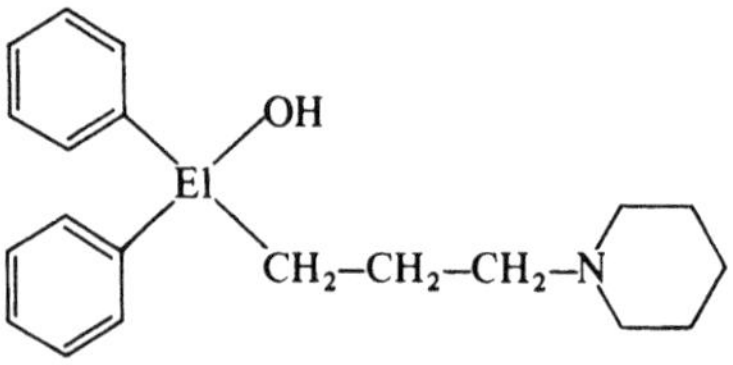

173a: El = C
173b: El = Si

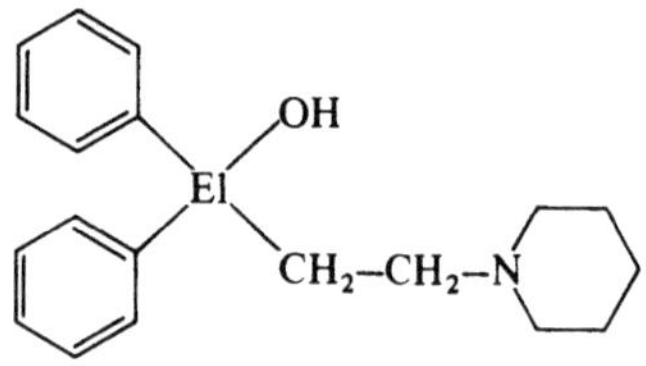

174a: El = C
174b: El = Si

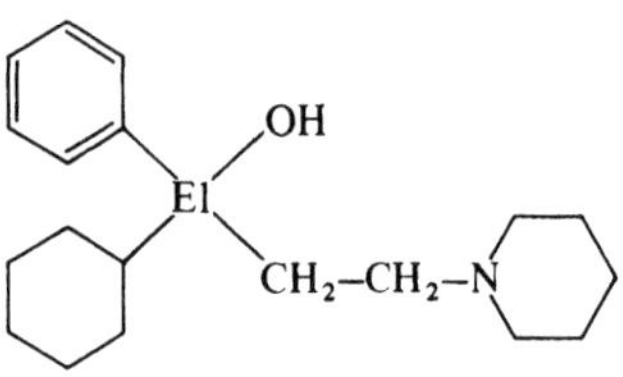

175a: El = C
175b: El = Si

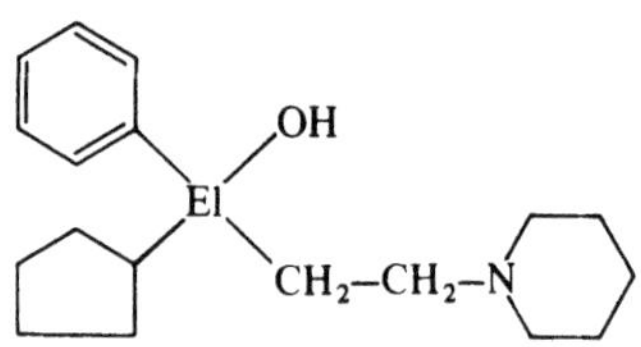

176a: El = C
176b: El = Si

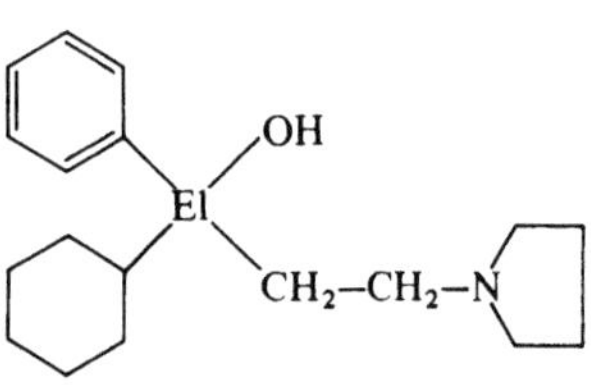

177a: El = C
177b: El = Si

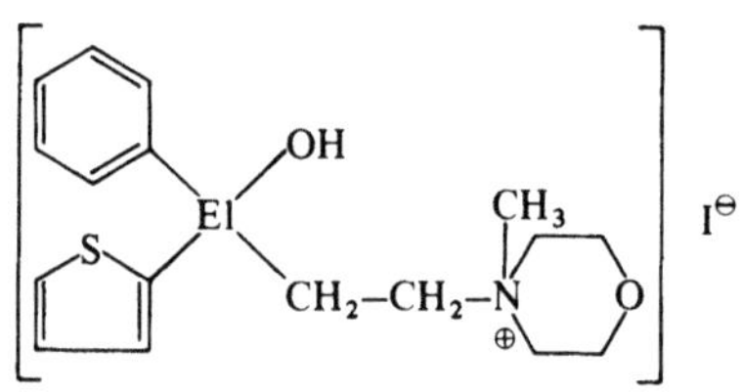

178a: El = C
178b: El = Si

totype, difenidol (*173a*), is used as an antiemetic. The lower homologue, pridinol (*174a*), is effective chiefly as a central muscle relaxant and as antiparkinson agent. Trihexyphenidyl (*175a*), cycrimine (*176a*), and procyclidine (*177a*) are antispasmodic agents, which have found some use in the treatment of the symptoms of Parkinson's disease. Tiemonium iodide (*178a*), a quaternary ammonium derivative of *179a*, is also a potent spasmolytic. The central OH group of all these molecules seems to be very important for the anticholinergic component. Not only substitution of the OH group by a hydrogen atom or a methyl group but also its esterification and etherification leads to a decreased anticholinergic activity.

However, replacement of the OH group by a $CO{-}NH_2$ group does not decrease strongly the parasympatholytic component. The great importance of the OH (and $CO{-}NH_2$) group seems to be related to the acidic hydrogen atom, which could interact with the corresponding receptor by a hydrogen bridge linkage.

Silanols exhibit normally a greater O–H acidity than analogous carbinols (cf. Chap. 2). This leads to the hypothetical conclusion that the sila-analogues *173b–178b* should exhibit a stronger affinity to the receptor than the corresponding carbon compounds *173a–178a.*

Starting with this hypothesis, syntheses[116, 119, 122–124] and comparative pharmacological-toxicological investigations[112, 120, 121] of the sila-analogues sila-difenidol (*173b*), sila-pridinol (*174b*), sila-trihexyphenidyl (*175b*), sila-cycrimine (*176b*), sila-procyclidine (*177b*), sila-tiemonium iodide (*178b*), and some of their derivatives (*179b, 180–183*) were carried out. The syntheses of the pairs difenidol/sila-difenidol and tiemonium iodide/sila-tiemonium iodide are described in Scheme 22 and Scheme 23, respectively.

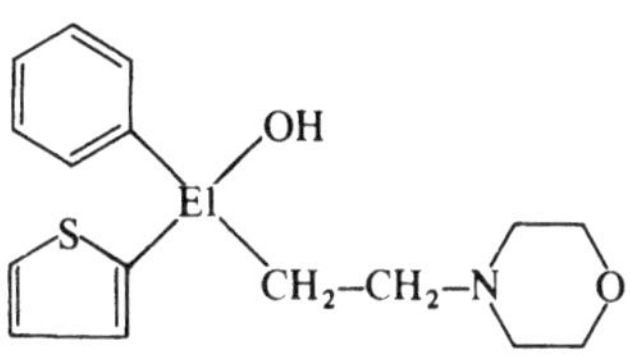

179a: El = C
179b: El = Si

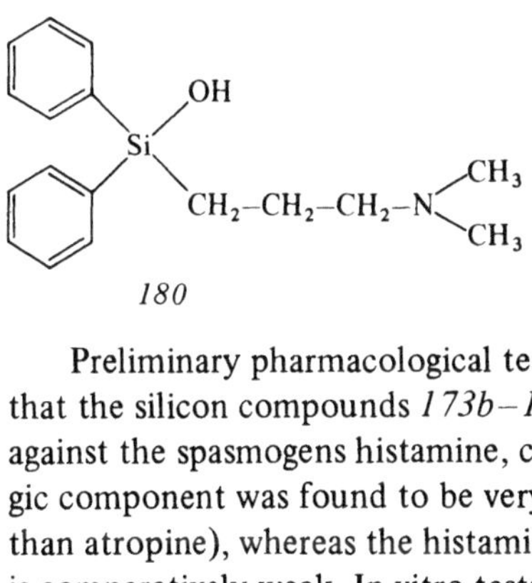

180

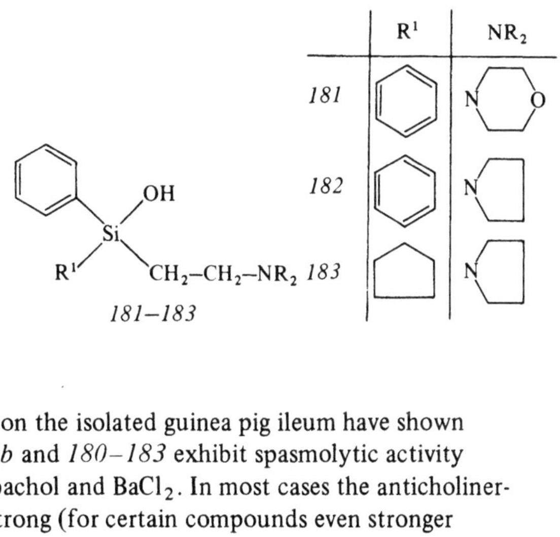

181–183

Preliminary pharmacological tests on the isolated guinea pig ileum have shown that the silicon compounds *173b–179b* and *180–183* exhibit spasmolytic activity against the spasmogens histamine, carbachol and $BaCl_2$. In most cases the anticholinergic component was found to be very strong (for certain compounds even stronger than atropine), whereas the histaminolytic and musculotropic spasmolytic activity is comparatively weak. In vitro tests with the carbon compounds *173a, 174a, 178a* and *179a* and their sila-analogues (*173b, 174b, 178b* and *179b*) have shown that the silicon compounds exhibit either the same or even a stronger activity against the spasmogens histamine and carbachol (compare Table 17). Sila-difenidol and *179b* were found to be more active against carbachol than their carbon analogues *173a* and *179a,* respectively: sila-difenidol is about 20 times more active than difenidol, and *179b* acts 7 times stronger than *179a.* The silicon compound *179b* exhibits also a greater histaminolytic activity (~5 : 1) than its carbon analogue *179a.* Sila-tiemonium iodide was found to be 6 times more active against histamine than tiemonium iodide itself.

In vitro tests with the pair *173a*/*173b* on the left auricle of guinea pigs resulted also in different intensities of activity. Both compounds were found to lead to a prolongation of the refractory period. However, sila-difenidol exhibits a significant

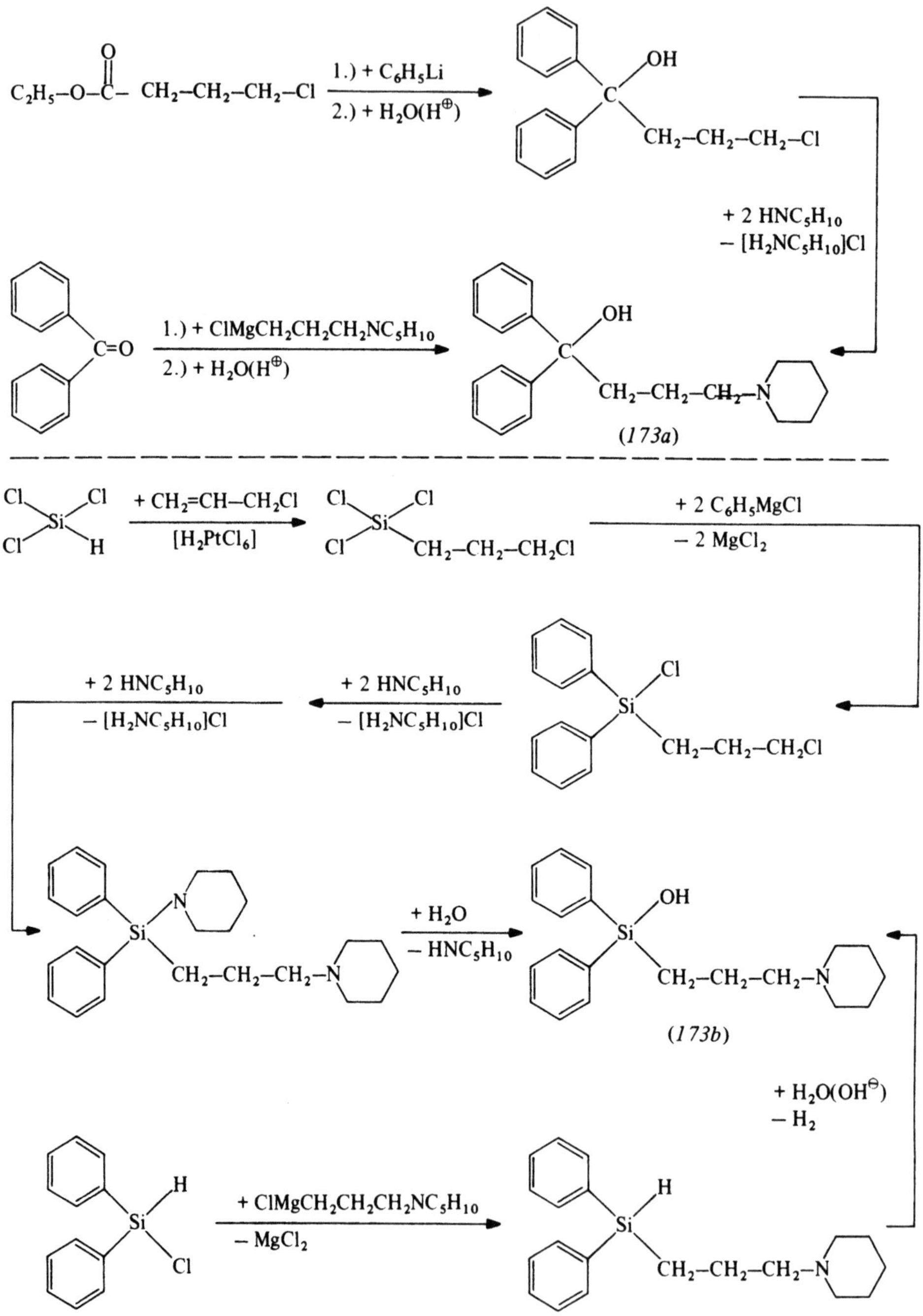

Scheme 22. Syntheses of difenidol and sila-difenidol

larger intensity as compared with its carbon analogue. The results of toxicological investigations with the pairs *173a*/*173b* and *174a*/*174b* are listed in Table 18. Sila-pridinol was found to be more toxic than its carbon analogue, whereas difenidol and sila-difendiol exhibit the same toxicity.

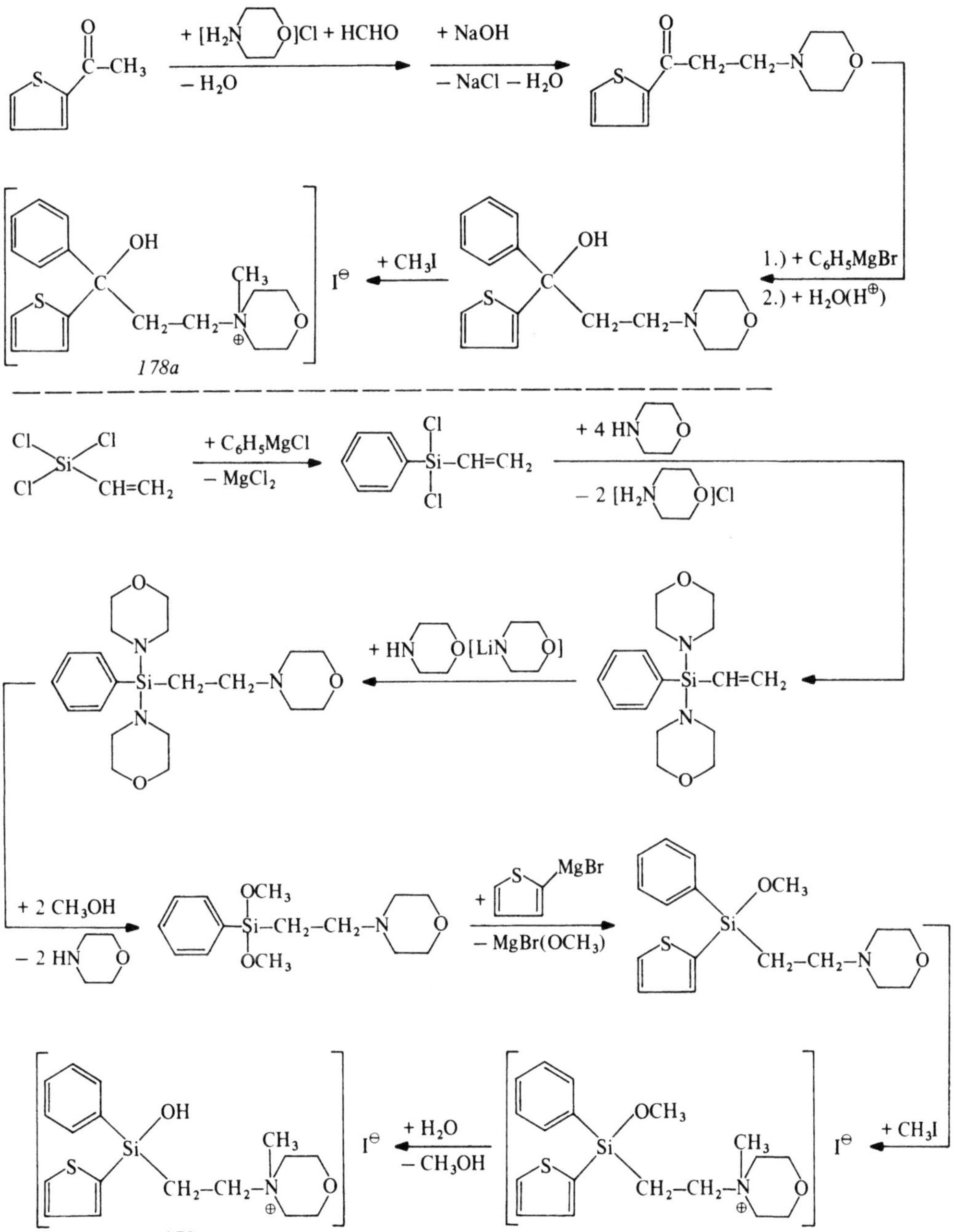

Scheme 23. Syntheses of tiemonium iodide and sila-tiemonium iodide

Table 17. Spasmolytic activities[120] of pairs *173a*/*173b*, *174a*/*174b*, *178a*/*178b*, and *179a*/*179b*

Compound	ED_{50}[mol/l][a] histamine	ED_{50}[mol/l][a] carbachol	ED_{50}[mol/l][a] $BaCl_2$
173a	5.9×10^{-7} (1.8×10^{-7}–19.9×10^{-7})	7.4×10^{-9} (2.3×10^{-9}–23.4×10^{-9})	–
173b	2.5×10^{-7} (0.9×10^{-7}–6.9×10^{-7})	3.5×10^{-10} (1.0×10^{-10}–12.5×10^{-10})	2.3×10^{-5} (0.6×10^{-5}–9.1×10^{-5})
174a	4.5×10^{-7} (1.3×10^{-7}–15.2×10^{-7})	1.6×10^{-9} (0.4×10^{-9}–5.6×10^{-9})	6.2×10^{-5} (0.8×10^{-5}–13.8×10^{-5})
174b	5.1×10^{-7} (1.5×10^{-7}–17.5×10^{-7})	0.5×10^{-9} (0.1×10^{-9}–1.9×10^{-9})	1.6×10^{-5} (0.3×10^{-5}–11.0×10^{-5})
178a	3.5×10^{-6} (0.6×10^{-6}–20.5×10^{-6})	1.3×10^{-9} (0.3×10^{-9}–6.4×10^{-9})	1.5×10^{-5} (0.3×10^{-5}–7.2×10^{-5})
178b	5.6×10^{-7} (1.3×10^{-7}–24.0×10^{-7})	1.3×10^{-9} (0.2×10^{-9}–6.2×10^{-9})	8.1×10^{-5} (0.2×10^{-5}–28.1×10^{-5})
179a	1.3×10^{-6} (0.4×10^{-6}–5.1×10^{-6})	4.3×10^{-8} (0.8×10^{-8}–24.0×10^{-8})	1.5×10^{-5} (0.2×10^{-5}–8.5×10^{-5})
179b	2.5×10^{-7} (0.9×10^{-7}–10.2×10^{-7})	6.3×10^{-9} (2.2×10^{-9}–18.0×10^{-9})	2.2×10^{-5} (0.6×10^{-5}–7.3×10^{-5})

[a] Confidence limits (95%) in brackets.

Table 18. Acute toxicities[112, 121] (white mice, i.p.) of pairs *173a*/*173b* and *174a*/*174b*

Compound	*173a*	*173b*	*174a*	*174b*
LD_{50}[mg/kg][a]	103.0 (95.8–110.8)	101.6 (94.5–109.3)	104.5 (98.4–111.1)	65.8 (59.2–73.2)

[a] Confidence limits (95%) in brackets.

6.4.2 Silicon Containing Derivatives of Tertiary Alcohols with Sedative-Hypnotic Activity

The pharmacological comparison[60] of diphenylmethylsilanol (*184b*) and its carbon analogue *184a* as anticonvulsants and sedatives resulted in quantitative and qualitative differences. *184b* was found to exhibit a stronger anticonvulsant potency than the carbon analogue (electroshock test, compounds were given orally to male mice; *184a:* ED_{50} = 154 mg/kg; *184b:* ED_{50} = 80 mg/kg). However, in contrast to *184a* the silanol *184b* does not lead to a sedation at anticonvulsant dose levels.

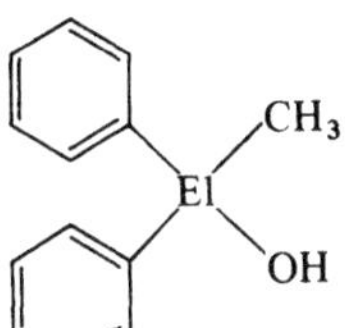

184a: El = C
184b: El = Si

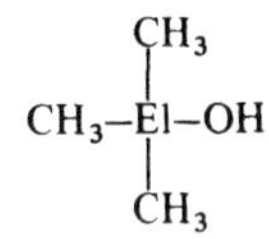

185a: El = C
185b: El = Si

R^1, R^2 = alkyl

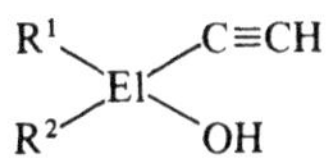

186a: El = C
186b: El = Si

The converse effect is seen in the comparison[60] of tert-butanol (*185a*) and the sila-analogue trimethylsilanol (*185b*); the ED_{50} values for sedation (loss of righting reflex in mice for 1 h) were found to be 1800 mg/kg for *185a* and 390 mg/kg for the silicon compound *185b*. Comparative investigations with hypnotics of type *186a* and their sila-analogues *186b* are in progress[125].

6.5 Comparison of Structures R^1–CH_2–R^2 and R^1–$Si(CH_3)_2$–R^2

Many of the drugs which are used medically contain one or more CH_2 groups. Bioactive sila-analogues with SiH_2 groups have not been described in the literature so far, possibly because of the hydrolytic sensitivity of Si–H bonds in the physiological environment. However, several attempts have been made to substitute CH_2 groups in biotropic structures for stable $Si(CH_3)_2$ groups.

6.5.1 Si,Si-Dimethyl-Sila-Substituted Derivatives of Spirobarbiturates with Narcotic Activity

Spirobarbiturates of type *187* show narcotic activity similar to the well known 5,5-dialkylbarbiturates, which are widely used as drugs. Fessenden et al.[126] report the synthesis and biological properties of the silicon containing compounds *190–193*, which can be regarded as sila-substituted derivatives of *187–189*, containing a $Si(CH_3)_2$ group instead of a CH_2 group. Compounds *194* and *195* with a seven-membered silicon containing ring were also prepared. In contrast to the carbon compounds of type *187*, the R-substituted silicon compounds *191* and *192* are readily to prepare. As an example, the synthesis of *191* is described in Scheme 24.

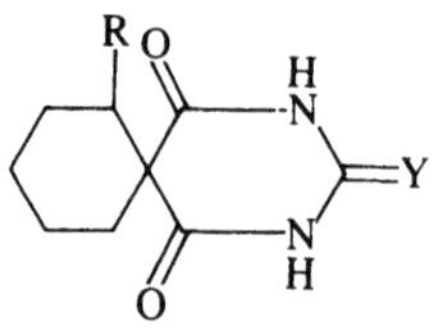

187–189

187: R = alkyl; Y = O, S
188: R = H, Y = O
189: R = H, Y = S

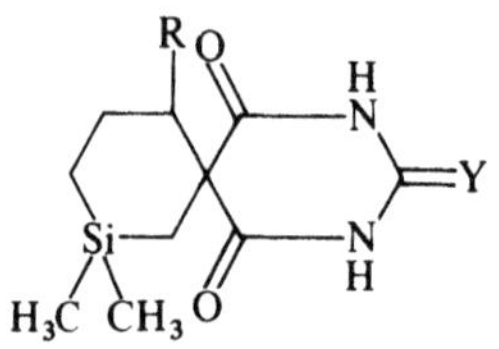

190–193

190: R = H, Y = O
191: R = CH_3, Y = O
192: R = C_2H_5, Y = O
193: R = H, Y = S

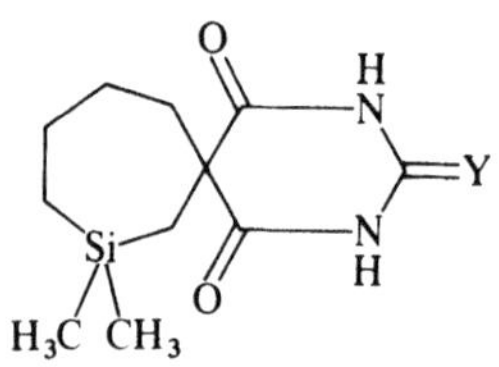

194, 195

194: Y = O
195: Y = S

The toxicity (LD_{50}) and the loss of the righting reflex (ED_{50}) have been evaluated for *188–195* (Table 19). In some cases the silicon compounds *190–195* were found to exhibit a similar biological activity to the carbon compounds *188* and *189*. However, differences have also been found (e.g. different ratios of the effective doses for the loss of the righting reflex to the toxic doses: ED_{50}/LD_{50}). A clear explanation for these results cannot be given because data of direct carbon analogues of *190–195* with geminal methyl groups are not available so far.

Table 19. Pharmacological and toxicological effects of *188–195* after i.p. dosing to white mice [126)]

Compound	ED_{50}[mg/kg][a]	LD_{50}[mg/kg][b]
188	[c]	1000
189	[d]	600 (440–830)
190	670 (500–880)	1000
191	190 (135–270)	230 (165–320)
192	240 (185–320)	450 (330–530)
193	140 (100–200)	230 (165–320)
194	230 (165–320)	670 (500–880)
195	[d]	140 (100–200)

[a] Loss of righting reflex (95% confidence levels).
[b] Toxicity (95% confidence levels).
[c] No activity noted at 1000 mg/kg.
[d] All animals who lost the righting reflex died.

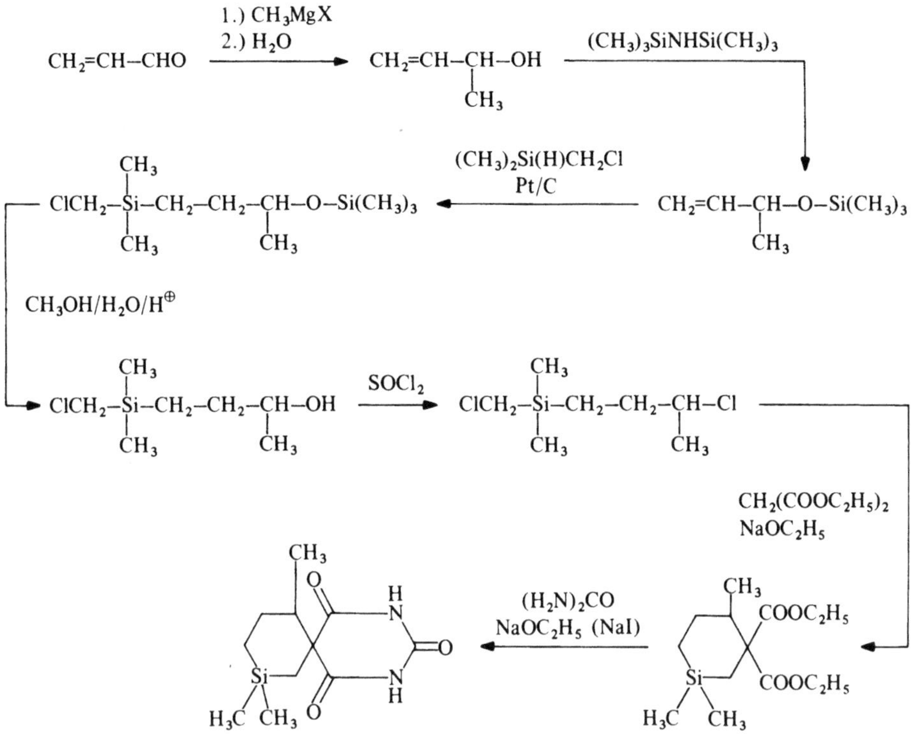

191

Scheme 24. Syntheses of the silicon containing spirobarbiturate *191*

6.5.2 Si,Si-Dimethyl-Sila-Substituted Derivatives of Oestradiol and Mestranol

According to Barcza[127], the skeleton of 6,6-dimethyl-6-sila-steroids could have interesting properties because the steroid 6-hydroxylation pathway would be blocked in such structures, and aromatization of ring B would be made impossible because the formation of a stable Si–C double bond is not possible. The sila-substitution in the middle of a polycondensed ring skeleton therefore prevents the eventual formation of polycondensed aromaticity avoiding possible carcinogenic effects.

Starting with this hypothesis, several 6-sila-steroids were prepared[127–129]. As an example, the syntheses of 6,6-dimethyl-6-sila-oestradiol (*197*) and 6,6-dimethyl-6-sila-mestranol (*198*) are described in Scheme 25. Not only the synthesis of the key intermediate 4,4-dimethyl-4-sila-6-methoxy-1-tetralone (*196*) but also the well known organic reactions, leading from *196* to *197* and *198*, are strongly influenced by the specific properties of the silicon atom. Although known reactions of steroid chemistry could be used for further transformations from *196* to *197* and *198*, conditions were not directly transferable because of the chemical behaviour of the $Si–C_{alkyl}$ and $Si–C_{aryl}$ bond. Important differences in the pH- and solvent profile were necessary. Nevertheless Scheme 25 shows very clearly that complicated organic chemistry is possible in the field of sila-pharmaca, even if one starts with silicon containing precursors.

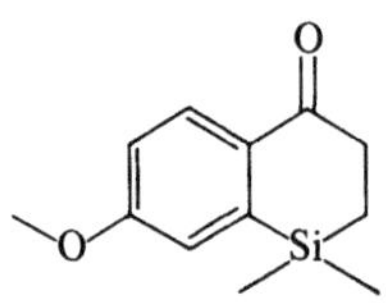

196

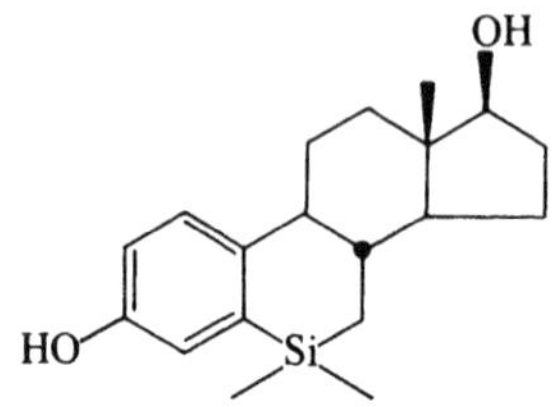

197

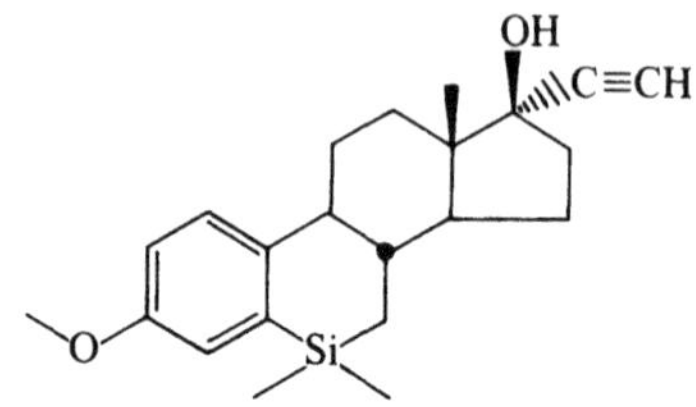

198

However, the biological evaluation[129)] of the 6,6-dimethyl-6-sila-steroids obtained by a great synthetic effort led to disappointing results. Compounds *197–207* were screened for oestrogenic, anti-oestrogenic, and postcoital activity. No significant oestrogenic or anti-oestrogenic activity was observed using doses 10^2-10^3 times that of an oestradiol standard. Compounds *197* and *206* showed post-coital activity in the rat, but only at 10 mg/kg, while *207* exhibited only a weak androgenic activity.

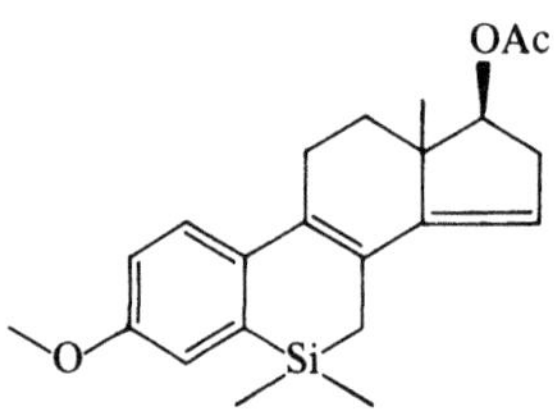

199

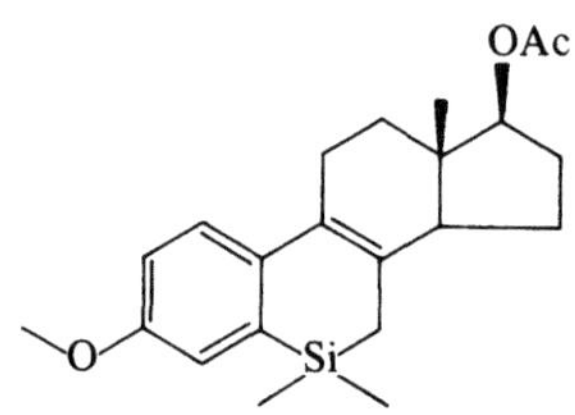

200

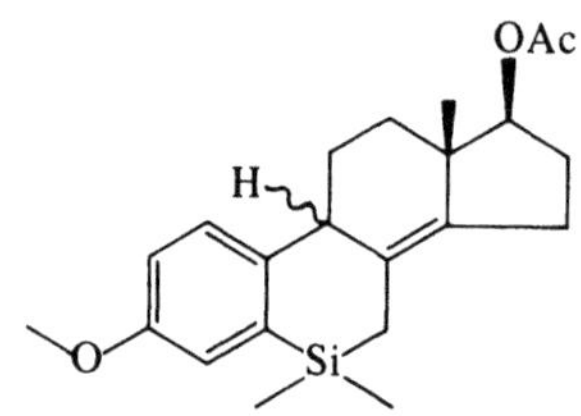

201

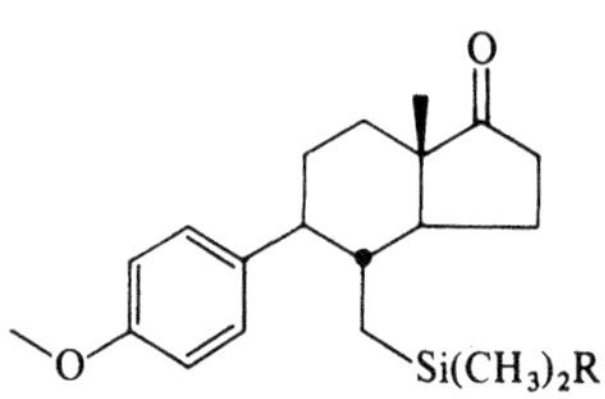

202: R = OH
203: R = F

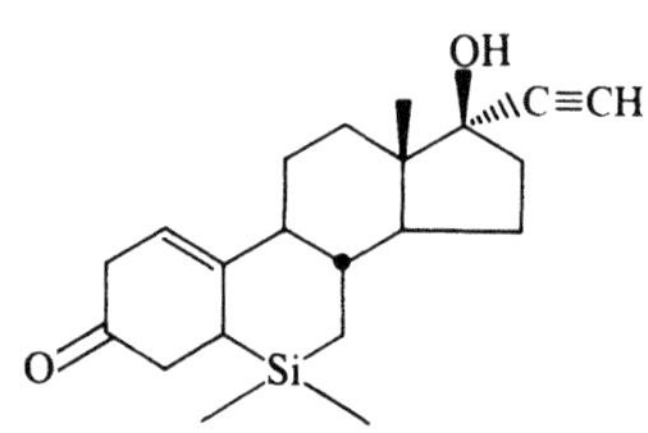

204

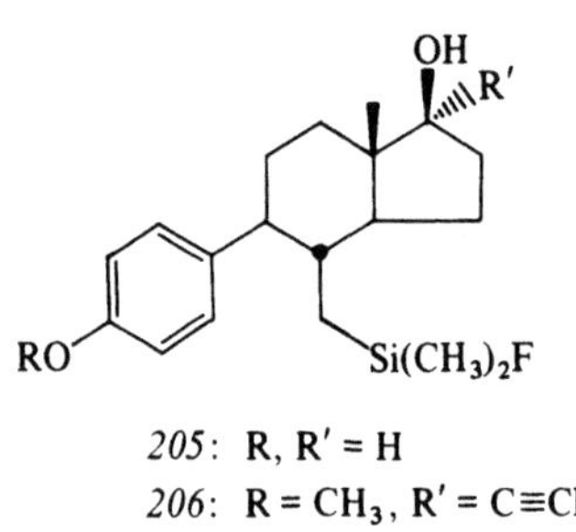

205: R, R′ = H
206: R = CH_3, R′ = C≡CH

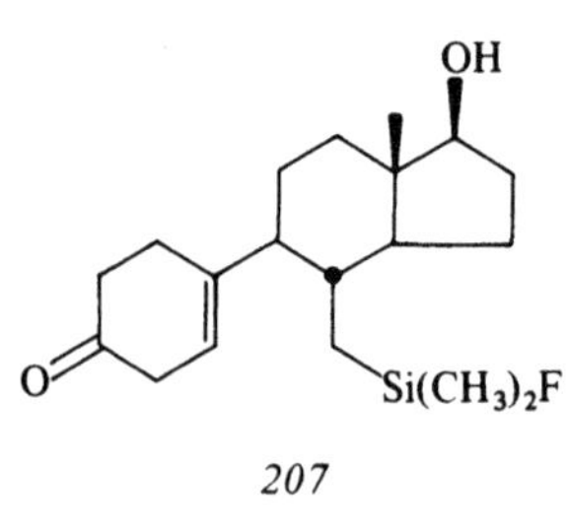

207

This lack of significant hormonal activity might be attributed to a metabolic effect, to a steric effect (steric inhibition of interaction with the uterine receptor protein(s), due either to the geminal dimethyl substituents at the silicon atom or to conformational changes in the total steroid skeleton resulting from the longer Si–C bond length in ring B), or to an electronic effect, generated by the specific properties of the silicon atom (Chap. 2).

However, the binding affinities of *197*, *206* and *207* [0.3%, 0.1% and <0.01%, respectively, relative to oestradiol (100%)] for the oestrogen specific acceptor protein of the rat uterus suggest that a steric or an electronic effect, rather than metabolic instability, is responsible for the absence of oestrogenic activity in these compounds. But a clear answer cannot be given because biological data of the direct 6,6-dimethyl-6-carba-analogues are not known so far.

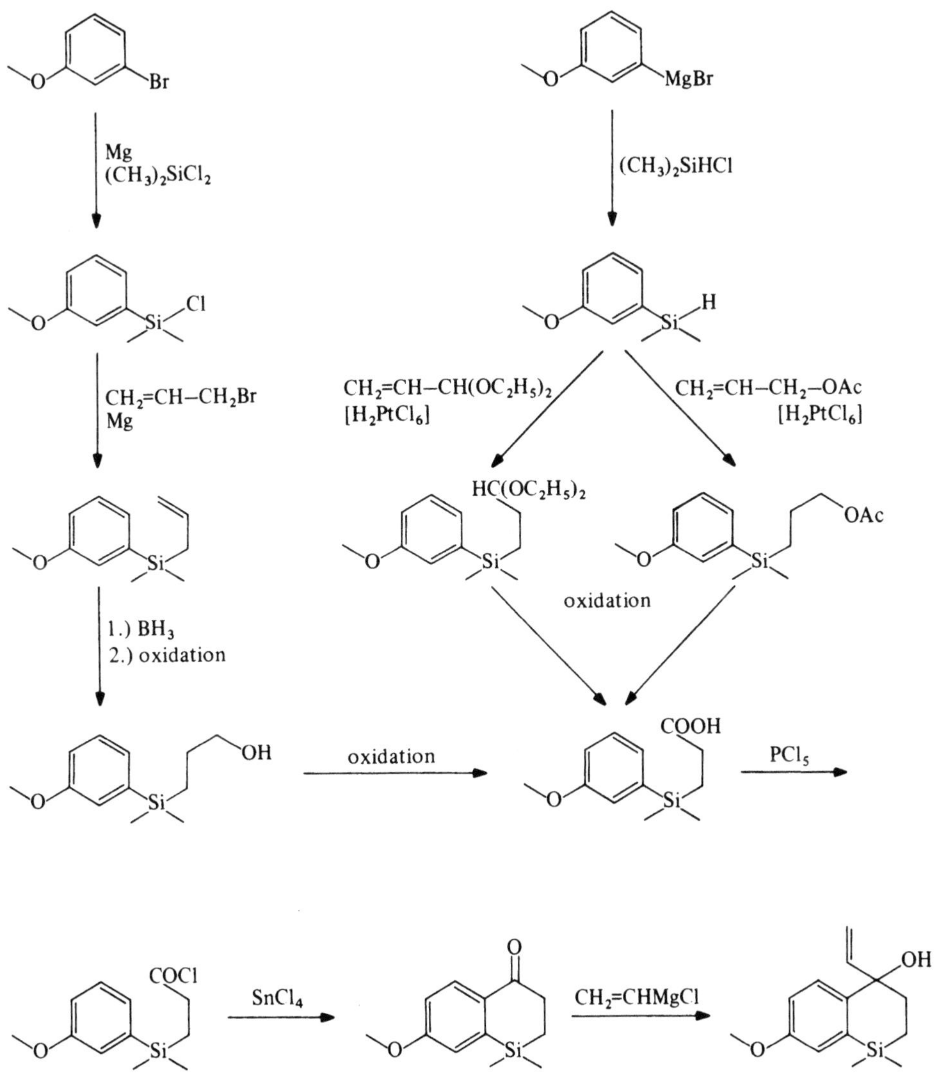

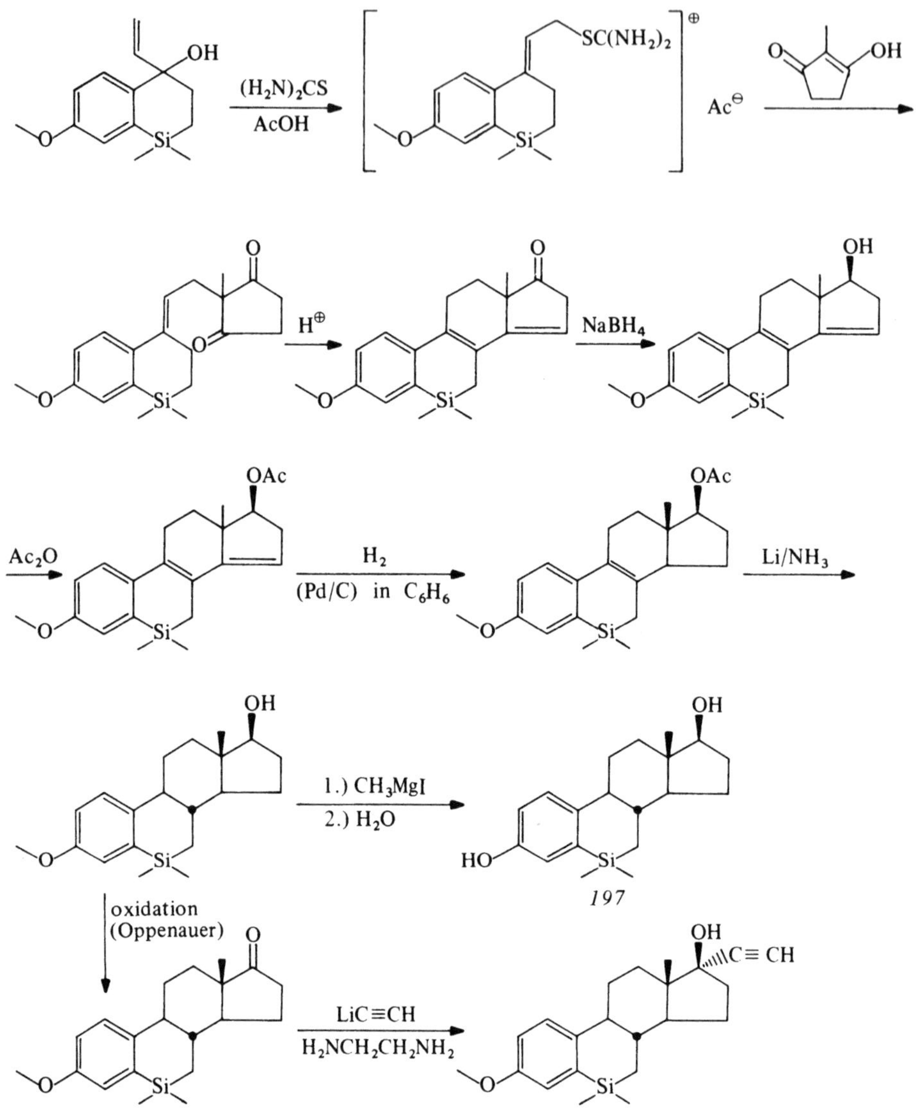

197

198

Scheme 25. Syntheses of 6,6-dimethyl-6-sila-oestradiol and 6,6-dimethyl-6-sila-mestranol

6.5.3 Si,Si-Dimethyl-Sila-Substituted Derivatives of Polymethylen-bis-Trimethyl-ammonium Compounds with Curare-Like Activity

Neuromuscular blocking agents (curare alkaloids and certain synthetic compounds), which interfere with transmission from motor nerve endings to the membrane of the skeletal muscle, are of great practical interest. They are used in anaesthesia (to reduce the muscle tonus) and in psychiatric electroshock therapy (to reduce the intensity of the accompanying muscular contractions). However, these agents are also of interest from a more theoretical point of view, because the neuromuscular junction serves as a good experimental model for transmission between cells and for studying drug-receptor interactions.

Although there is a considerable variation in overall structure of most neuromuscular blocking agents, there is one structural feature which is typical: two quaternary ammonium groups are separated by a certain distance. In general it has been found that bisquaternary ammonium compounds are most active if the nitrogen atoms are about 14 Å apart. The occurrence of a sharp maximum of curare-like activity in the polymethylene-bis-trimethylammonium series $(CH_3)_3N(CH_2)_nN(CH_3)_3^{\oplus\oplus}$ for the decamethylene compound (n = 10) is a good example of this structure-activity relationship.

In the synthetic drugs decamethonium (*208*) and suxamethonium (*209*) the 14 Å distance is realized by a skeleton of 10 carbon atoms, and by a chain of 8 carbon atoms and 2 oxygen atoms, respectively.

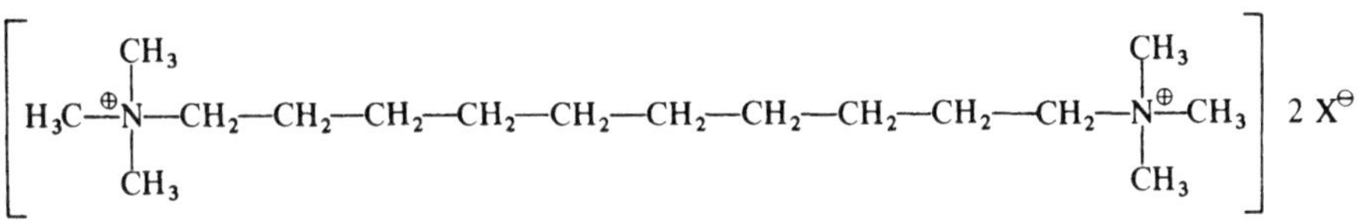

208

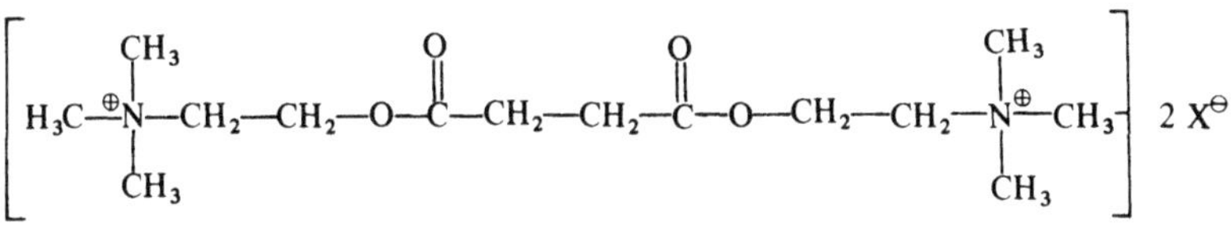

209

According to the different covalent radii of carbon and silicon (Chap. 2), the exchange of the C_{10} chain of decamethonium for a 9 atom chain of 7 carbon atoms and 2 silicon atoms with a $C_3SiCSiC_3$ sequence should lead to a similar N–N distance. Supposing that all CCC- and CSiC-bond angles are equal (tetrahedral angles), and

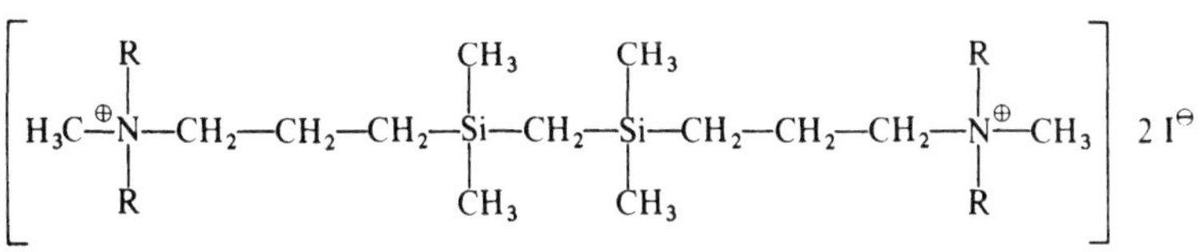

210: R = CH_3
211: R = C_2H_5

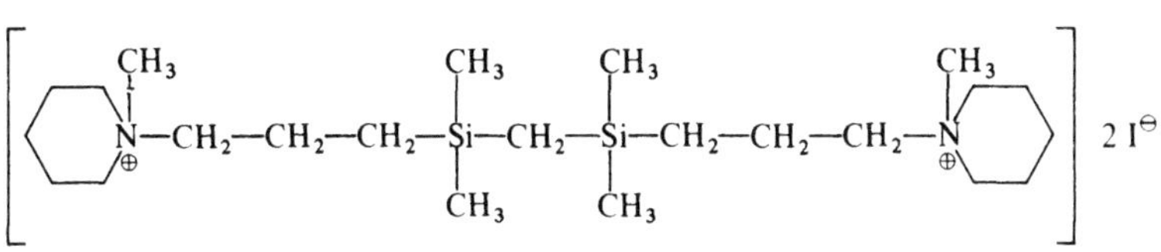

212

supposing that the 9 atoms of the C_7Si_2 skeleton and the 2 nitrogen atoms of the ammonium groups are lying in the same plane, the nitrogen atoms in compounds *210–212* would be separated by a distance of about 14 Å, which should result in a curare-like activity.

A C_6Si_2O-skeleton with a $C_3SiOSiC_3$ sequence should also lead to a similar N–N distance. Only a little shortening is expected because of the Si–O bond length and SiOSi bond angle. According to this concept, the disiloxanes *213–215* should also exhibit curare-like activity.

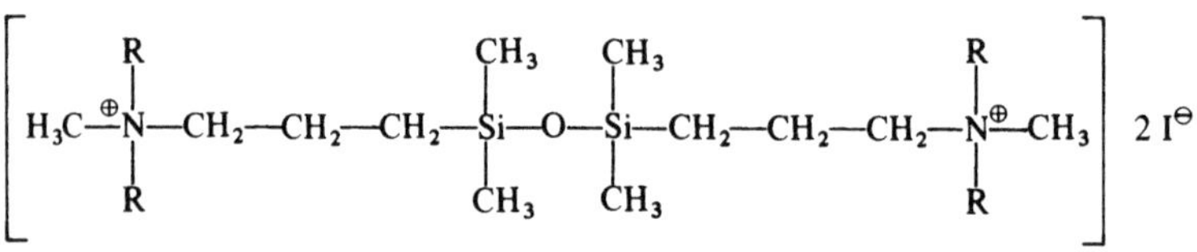

213: R = CH_3
214: R = C_2H_5

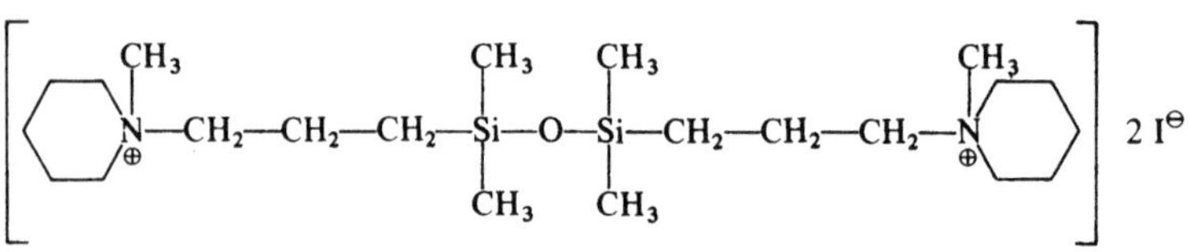

215

Starting with this hypothetical structure-activity relationship, the syntheses and biological investigations of *210–215* and decamethonium iodide were carried out[130, 131]. As an example, the synthesis of *212* is described in Scheme 26. The pharmacological and toxicological data are given in Table 20 and Table 21, respectively.

In the mice head-drop test for curare-like activity compounds *210, 212* and *213* were found to be very potent. In analogy to d-tubocurarine, decamethonium and other muscle relaxants, these compounds cause a neuromuscular block when administered intraperitoneally.

Overdoses of *210–215* lead to death because of respiratory paralysis. The LD_{50} values are very low (see Table 21) and all compounds are even more toxic than the standard decamethonium iodide. In addition, the most toxic agent *212*, which is nearly 30 times more toxic than the standard decamethonium iodide, seems to be the most toxic organosilicon compound described in the literature up to now.

Table 20. Pharmacological data of *210, 212* and *213*[130]

Compound	ED_{50}[mg/kg][a]	ED_{50}[μmol/kg][a]
210	0.29 (0.20–0.43)	0.50 (0.34–0.73)
212	0.07 (0.05–0.10)	0.11 (0.08–0.16)
213	0.80 (0.50–1.18)	1.35 (0.85–2.01)

[a] Head drop after i.p. application to white mice, confidence limits (95%) in brackets.

Table 21. Toxicological data of *210–215* and decamethonium iodide[130, 131]

Compound	LD_{50}[mg/kg][a]	LD_{50}[μmol/kg][a]
210	1.14 (0.99–1.31)	1.94 (1.69– 2.23)
211	1.03 (0.66–1.61)	1.60 (1.02– 2.51)
212	0.32 (0.28–0.37)	0.48 (0.42– 0.55)
213	1.91 (1.63–2.24)	3.25 (2.77– 3.81)
214	0.44 (0.32–0.61)	0.68 (0.49– 0.95)
215	0.76 (0.65–0.89)	1.14 (0.97– 1.33)
Decamethonium iodide	7.28 (6.62–8.01)	14.21 (12.92–15.64)

[a] Acute toxicity after i.p. administration to white mice, confidence limits (95%) in brackets.

In this connection it shall be pointed out that some permethylated α,ω-bis-(ammoniummethyl)polysiloxanes of type *216* are also very toxic, depending on the number of SiO-units[132]. This could possibly be explained by the structural similarity to decamethonium.

$$\left[R_3\overset{\oplus}{N}CH_2[Si(CH_3)_2O]_nSi(CH_3)_2CH_2\overset{\oplus}{N}R_3\right]2\,X^{\ominus} \qquad \textit{216}$$

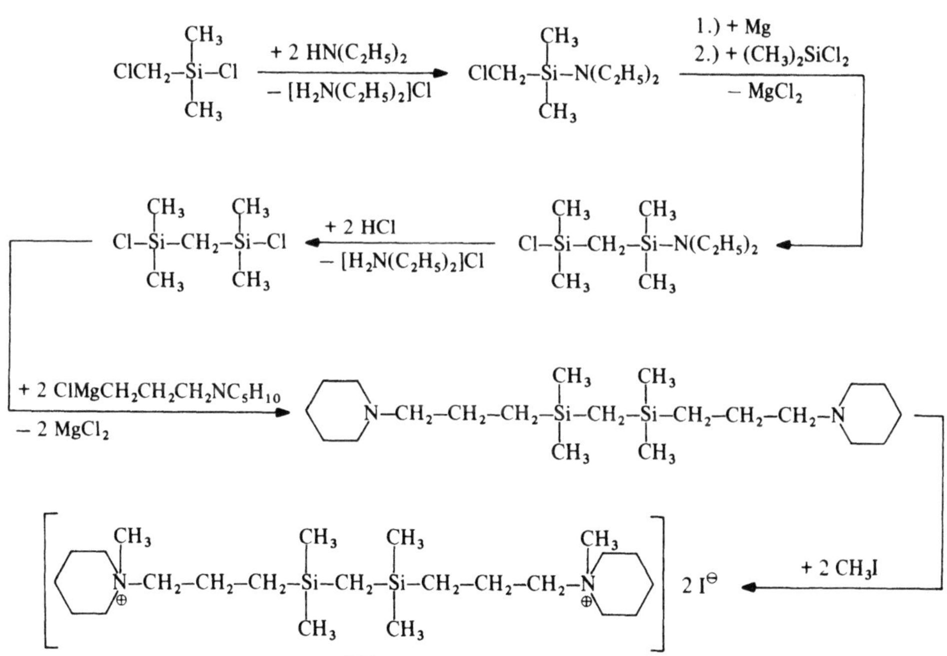

213

Scheme 26. Synthesis of the curare-like organosilicon agent *212*

7 References

1. Fessenden, R. J., Fessenden, J. S.: Advan. Drug Res. *4*, 95 (1967)
2. Voronkov, M. G.: Pure Appl. Chem. *19*, 399 (1969)
3. Voronkov, M. G., Lukevics, E.: Russ. Chem. Rev. *38*, 975 (1969)
4. Garson, L. R., Kirchner, L. K.: J. Pharm. Sci. *60*, 1113 (1971)
5. Bien, E.: Pharmazie *26*, 577 (1971)
6. Voronkov, M. G.: XXIVth Intern. Congress of Pure Appl. Chem., Vol. IV, pp. 45–66. London: Butterworth, 1974
7. Voronkov, M. G.: Chemistry in Britain *9*, 411 (1973)
8. Thayer, J. S.: J. Organometal. Chem. *76*, 272–275 (1974)
9. Voronkov, M. G.: Annual reports in medicinal chemistry, Vol. X, pp. 265–273. New York: Academic Press, 1975
10. Voronkov, M. G., Zelchan, G. I., Lukevitz, E.: Silizium und Leben. Berlin: Akademie-Verlag, 1975.
11. Bendz, G., Lindquist, I. (eds.): Biochemistry of silicon and related problems. New York: Plenum Press, 1978
12. Bažant, V., Chvalovský, V., Rathouský, J.: Organosilicon compounds, Vol. I. Prague: Publishing House of the Czechoslovak Academy of Sciences, and New York: Academic Press, 1965
13. Bürger, H.: Fortschr. chem. Forsch. *9*, 1 (1967), and references therein
14. Noll, W.: Chemie und Technologie der Silicone. Weinheim: Verlag Chemie, 1968
15. Ebsworth, E. A. V.: Physical basis of the chemistry of the group IV elements. In: The bond to carbon. MacDiarmid, A. G. (ed.). New York: Marcel Dekker, 1968
16. Eaborn, C.: Organosilicon compounds. London: Butterworth, 1960
17. Sommer, L. H.: Stereochemistry, mechanism and silicon. New York: McGraw-Hill, 1965
18. Petrov, A. D., Mironov, B. F., Ponomarenko, V. A., Chernyshev, E. A.: Synthesis of organosilicon monomers. London: Heywood & Company Ltd., 1964
19. Bürger, H.: Angew. Chem. *85*, 519 (1973)
20. Glidewell, C.: Inorg. Chim. Acta Rev. *7*, 69 (1973)
21. Rochow, E. G.: The chemistry of silicon. In: Pergamon texts in inorganic chemistry. Vol. IX, chapter 15 of Comprehensive inorganic chemistry. Oxford: Pergamon Press, 1975
22. Bažant, V., Horák, M., Chvalovský, V., Schraml, J.: Handbook of organosilicon compounds, Vol. I. New York: Marcel Dekker, 1973
23. Holleman, A. F., Wiberg, E.: Lehrbuch der Anorganischen Chemie. Berlin: Walter de Gruyter, 1976
24. Cotton, F. A., Wilkinson, G.: Advanced inorganic chemistry, p. 112. New York: Interscience Publishers, 1972
25. Handbook of chemistry and physics, 52nd ed.. Cleveland, Ohio: The Chemical Rubber Company, 1971–1972
26. Johnson, A. D.: Some thermodynamic aspects of inorganic chemistry, pp. 158–160. Cambridge: University Press, 1968
27. Jarvie, A. W. P.: Organometal. Chem. Rev. A *6*, 153 (1970)
28. Pierce, A. E.: Silylation of organic compounds. Rockford, III: Pierce Chem. Co., 1968
29. Piekos, R., Kobylczyk, K., Osmialowski, K.: Sci. Pharm. *45*, 234 (1977)
30. Piekos, R., Teodorczyk, J.: Sci. Pharm. *43*, 246 (1975)
31. Piekos, R., Macholla, J., Sankowski, M., Stozkowska, W.: Sci. Pharm. *44*, 206 (1976)
32. Piekos, R., Teodorczyk, J.: Roczniki Chem. *49*, 1603 (1975)
33. Chang, E., Jain, V. K.: J. Med. Chem. *9*, 433 (1966)
34. Lukevics, E.: Biological activity of nitrogen-containing organosilicon compounds. In: Biochemistry of silicon and related problems, pp. 435–445. New York: Plenum Press, 1978
35. Frankel, M., Broze, M., Gertner, D., Rotman, A., Shenhar, A. Zilkha, A.: J. Med. Chem. *11*, 857 (1968)
36. Voronkov, M. G.: Pure Appl. Chem. *13*, 35 (1966)

37. Woronkow, M. G., Seltschan, G. I., Lapsina, A., Pestunowitsch, W. A.: Z.Chem. *8*, 214 (1968)
38. Voronkov, M. G.: Biological activity of silatranes. In: Biochemistry of silicon and related problems, pp. 395–433. New York: Plenum Press, 1978
39. Pestunovich, V. A., Voronkov, M. G., Sjdorkin, V. F., Shagun, V. A.: IVth Intern. Symposium on Organosilicon Chemistry, Moscow 1975, Abstracts, Vol. I, part 1, p. 190
40. Frye, C. L., Vogel, G. E., Hall, J. A.: J. Amer. Chem. Soc. *83*, 996 (1961)
41. Finestone, A. B.: U.S. Pat. 2.953.545 (1960); Chem. Abstr. *55*, 4045 (1961)
42. Finestone, A. B.: Ger. Pat. 1.131.681 (1962); Chem. Abstr. *58*, 4598 (1963)
43. Beiter, C. B., Schwarcz, M., Crabtree, G.: Soap Chem. Spec. *46*, 38 (1970)
44. Lukevics, E., Dremova, V. P., Simchenko, L. I., Voronkov, M. G.: Khim.-Farm Zh. *8*, 29 (1974); Chem. Abstr. *82*, 57805p (1975)
45. Lukevics, E., Zile, A., Kruzmetra, L. V., Khudobin, Yu. I., Voronkov, M. G.: Khim.-Farm. Zh. *9*, 14 (1975); Chem. Abstr. *82*, 119469t (1975)
46. Lukevics, E., Zile, A., Kruzmetra, L., Pestunovich, A. E., Voronkov, M. G.: Latv. PSR Zinat. Akad. Vestis *3*, 83 (1974); Chem. Abstr. *81*, 146280k (1974)
47. Lukevics, E., Voronkov, M. G.: Dokl. Akad. Nauk SSSR *216*, 103 (1974); Chem. Abstr. *81*, 86427b (1974)
48. Olson, K. J.: Toxicol. Appl. Pharmacol. *21*, 12 (1972)
49. Palazzolo, R. J., McHard, J. A., Hobbs, E. J., Fancher, O. E., Calandra, J. C.: Toxicol. Appl. Pharmacol. *21*, 15 (1972)
50. LeFevre, R., Coulston, F., Goldberg, L.: Toxicol. Appl. Pharmacol. *21*, 29 (1972)
51. Hobbs, E. J., Fancher, O. E., Calandra, J. C.: Toxicol. Appl. Pharmacol. *21*, 45 (1972)
52. Bennett, D. R., Gorzinski, S. J., LeBeau, J. E.: Toxicol. Appl. Pharmacol. *21*, 55 (1972)
53. Hayden, J. F., Barlow, S. A.: Toxicol. Appl. Pharmacol. *21*, 68 (1972)
54. LeVier, R. R., Jankowiak, M. E.: Toxicol. Appl. Pharmacol. *21*, 80 (1972)
55. LeVier, R. R., Jankowiak, M. E.: Biol. Reprod. *7*, 260 (1972)
56. Bennett, D. R., McHard, J. A.: U.S. Pat. 3.830.912 (1974); Chem. Abstr. *82*, 39019f (1975)
57. Bennett, D. R., LeVier, R. R.: U.S. Pat. 3.821.373 (1974); Chem. Abstr. *82*, 26147w (1975)
58. Strindberg, B.: Biochemical effects of 2,6-cis-diphenylhexamethylcyclotetrasiloxane in man. In: Biochemistry of silicon and related problems, pp. 515–520. New York: Plenum Press, 1978
59. Vessman, J., Hammar, C.-G., Lindeke, B., Strömberg, S., LeVier, R., Robinson, R., Spielvogel, D., Hanneman, L.: Analysis of some organosilicon compounds in biological material. In: Biochemistry of silicon and related problems, pp. 535–558. New York: Plenum Press, 1978
60. LeVier, R. R., Chandler, M. L., Wendel, S. R.: The pharmacology of silanes and siloxanes. In: Biochemistry of silicon and related problems, pp. 473–514. New York: Plenum Press, 1978
61. Hyde, J. F., Speilvogel, D. E.: Ger. Offen. 2.032.826 (1971)
62. Carlström, D.: Structural aspects on organosilicon compounds. In: Biochemistry of silicon and related problems, pp. 523–534. New York: Plenum Press, 1978
63. Kipping, F. S.: J. Chem. Soc. *101*, 2125 (1912)
64. Hulce, V. D., Rech, R. H.: Pharmacologist *16*, 228 (1974)
65. Hulce, V. D., Rech, R. H.: Pharmacologist *17*, 178 (1975)
66. Fawcett, J. K., Camerman, N.: Can. J. Chem. *55*, 3631 (1977)
67. Loeper, J., Loeper, J., Fragny, M.: The physiological role of the silicon and its antiatheromatous action. In: Biochemistry of silicon and related problems, pp. 281–296. New York: Plenum Press, 1978
68. LeVrier, M., Reboul, J., Dufaut, N., Dilhuydy: Senologia *2*, 3 (1977)
69. Fessenden, R. J., Coon, M. D.: J. Med. Chem. *8*, 604 (1965)
70. Fessenden, R. J., Coon, M. D.: J. Med. Chem. *9*, 262 (1966)
71. Fessenden, R. J., Rittenhouse, R.: J. Med. Chem. *11*, 1070 (1968)
72. Fessenden, R. J., Ahlfors, C.: J. Med. Chem. *10*, 810 (1967)
73. Fessenden, R. J., Hartman, R. A.: J. Med. Chem. *13*, 52 (1970)
74. Henderson, P. T., Ariëns, E. J., Ellenbroek, B. W. J., Simonis, A. M.: J. Pharm. Pharmacol. *20*, 26 (1968)
75. Kröning, G., Schulz, E., Sprung, W.-D.: Pharmazie *30*, 765 (1975)

76. Fessenden, R. J., Coon, M. D.: J. Med. Chem. *7*, 561 (1964)
77. Rice, L. M.: Ger. Offen. 2.243.550 (1973); Chem. Abstr. *79*, 42680h (1973)
78. Rice, L. M., Sheth, B. S., Wheeler, J. W.: J. Heterocyclic Chem. *10*, 731 (1973)
79. Rice, L. M., Sheth, B. S., Wheeler, J. W.: J. Heterocyclic Chem. *10*, 737 (1973)
80. Metcalf, R. L., Fukuto, T. R.: J. Econ. Entomol. *58*, 1151 (1965)
81. Tacke, R., Strecker, M., Niedner, R., Oettingen, U. v.: Technische Universität Braunschweig, unpublished results
82. Tacke, R., Haller, I., Zeiler, H.-J.: Technische Universität Braunschweig, unpublished results
83. Speier, J. L.: U.S. Pat. 2.645.630 (1953); Chem. Abstr. *47*, 11810 (1953)
84. Speier, J. L.: J. Amer. Chem. Soc. *74*, 1003 (1952)
85. Frisch, K. C., Shroff, P. D.: J. Amer. Chem. Soc. *75*, 1249 (1953)
86. Fregert, S., Rorsman, H.: Nature *192*, 989 (1961)
87. Speier, J. L.: U.S. Pat. 2.611.776 (1952)
88. Raubach, H., Wende, A.: Ger. Pat. 1.157.612 (1963); Chem. Abstr. *60*, 6868 (1964)
89. Davidsohn, W., Laliberte, B. R., Goddard, C. M., Henry, M. C.: J. Organometal. Chem. *36*, 283 (1972)
90. Tacke, R., Heeg, E., Berndt, B.: Arch. Pharm., in press
91. Stach, K., Pöldinger, W.: Fortschr. Arzneimittelforsch. *9*, 151 (1966)
92. Wilhelm, M.: Pharm. J. *214*, 414 (1975)
93. Wiese, D., Tacke, R., Thewalt, U., Wannagat, U.: Technische Universität Braunschweig, unpublished results
94. Wiese, D.: Dissertation, Technische Universität Braunschweig, 1978
95. Wannagat, U.: Sila-pharmaca. In: Biochemistry of silicon and related problems, pp. 447–471. New York: Plenum Press, 1978
96. Wannagat, U.: Jahrbuch der Akademie der Wissenschaften in Göttingen, pp. 78–104. Göttingen: Vandenhoeck & Ruprecht, 1976
97. Corey, J. Y., Dye, M. J., Farrell, R. L., Mitchell, M. V.: J. Organometal. Chem. *153*, 127 (1978)
98. Corey, J. Y., Paton, J. P., Rankin, D. M.: J. Organometal. Chem. *139*, 1 (1977)
99. Corey, J. Y., Farrell, R. L.: J. Organometal. Chem. *153*, 15 (1978)
100. Corey, E. R., Corey, J. Y., Glick, M. D.: Acta Cryst. *B32*, 2025 (1976)
101. Corey, E. R., Corey, J. Y., Paton, W. F.: Acta Cryst. *B33*, 1254 (1977)
102. Tacke, R.: Dissertation, Technische Universität Braunschweig, 1974
103. Tacke, R., Wannagat, U.: IVth Intern. Symposium on Organosilicon Chemistry, Moscow 1975, Abstracts, Vol. I, part 2, p. 3
104. Tacke, R., Wannagat, U.: Mh. Chem. *106*, 1005 (1975)
105. Tacke, R., Wannagat, U.: Mh. Chem. *107*, 111 (1976)
106. Tacke, R., Wannagat, U.: Mh. Chem. *107*, 439 (1976)
107. Tacke, R., Wannagat, U.: Mh. Chem. *107*, 449 (1976)
108. Tacke, R., Wannagat, U.: Mh. Chem. *107*, 1265 (1976)
109. Tacke, R., Wannagat, U.: Mh. Chem. *107*, 1271 (1976)
110. Tacke, R., Wannagat, U.: Arch. Pharm. *310*, 714 (1977)
111. Tacke, R.: Arch. Pharm. *310*, 719 (1977)
112. Rossée, H.-U.: Dissertation, Technische Universität Braunschweig, 1974
113. Kuhrt, G., Matthies, H., Liebmann, H., Rühlmann, K.: Pharmazie *31*, 849 (1976)
114. Friedrich, G., Bartsch, R., Rühlmann, K.: Pharmazie *32*, 394 (1977)
115. Ackermann, J.: Dissertation, Technische Universität Braunschweig, 1977
116. Ackermann, J., Steiling, L., Tacke, R., Wannagat, U.: IVth Intern. Symposium on Organosilicon Chemistry, Moscow 1975, Abstracts, Vol. I, part 1, p. 186
117. Ackermann, J., Tacke, R., Wannagat, U.: Technische Universität Braunschweig, unpublished results
118. Koke, U.: Dissertation, Technische Universität Braunschweig, 1977
119. Tacke, R., Strecker, M.: Technische Universität Braunschweig, unpublished results
120. Heeg, E., Berndt, B., Knapstein, C.-M.: Technische Universität Braunschweig, unpublished results

121. Niedner, R.: Technische Universität Braunschweig, unpublished results
122. Steiling, L., Tacke, R., Wannagat, U.: Technische Universität Braunschweig, unpublished results
123. Steiling, L.: Dissertation, Technische Universität Braunschweig, 1977
124. Tacke, R., Zimonyi-Hegedüs, E., Strecker, M.: Technische Universität Braunschweig, unpublished results
125. Tacke, R., Bentlage, A., Niedner, R.: Technische Universität Braunschweig, unpublisched results
126. Fessenden, R. J., Larsen, J. G., Coon, M. D., Fessenden, J. S.: J. Med. Chem. *7*, 695 (1964)
127. Barcza, S., Hoffman, C. W.: Tetrahedron *31*, 2363 (1975)
128. Barcza, S.: U. S. Pat. 3.637.782 (1972); Chem. Abstr. *76*, 154 021p (1972)
129. Pitt, C. G., Friedman, A. E., Rector, D., Wani, M. C.: Tetrahedron *31*, 2369 (1975)
130. Tacke, R., Niedner, R.: Z. Naturforsch. *33b*, 412 (1978)
131. Tacke, R., Niedner, R.: Technische Universität Braunschweig, unpublished results
132. Voronkov, M. G., Steiling, L., Kirpichenko, S. W., Kuznetsova, E. E.: 5th International Symposium on Organosilicon Chemistry, Karlsruhe 1978, Abstracts, p. 62

Received January 2, 1979

Biological Activity of Silatranes

Michail G. Voronkov

Institute of Organic Chemistry, Siberian Division of the USSR Academy of Sciences, 664033 Irkutsk, USSR

Table of Contents

Introduction

The high toxicity and specific biological activity of 1-arylsilatranes was discovered about 15 years ago[1, 2]. At that time, these substances belonged to an almost unknown class of heterocyclic derivatives of 5-coordinate silicon with general formula I (Fig. 1).

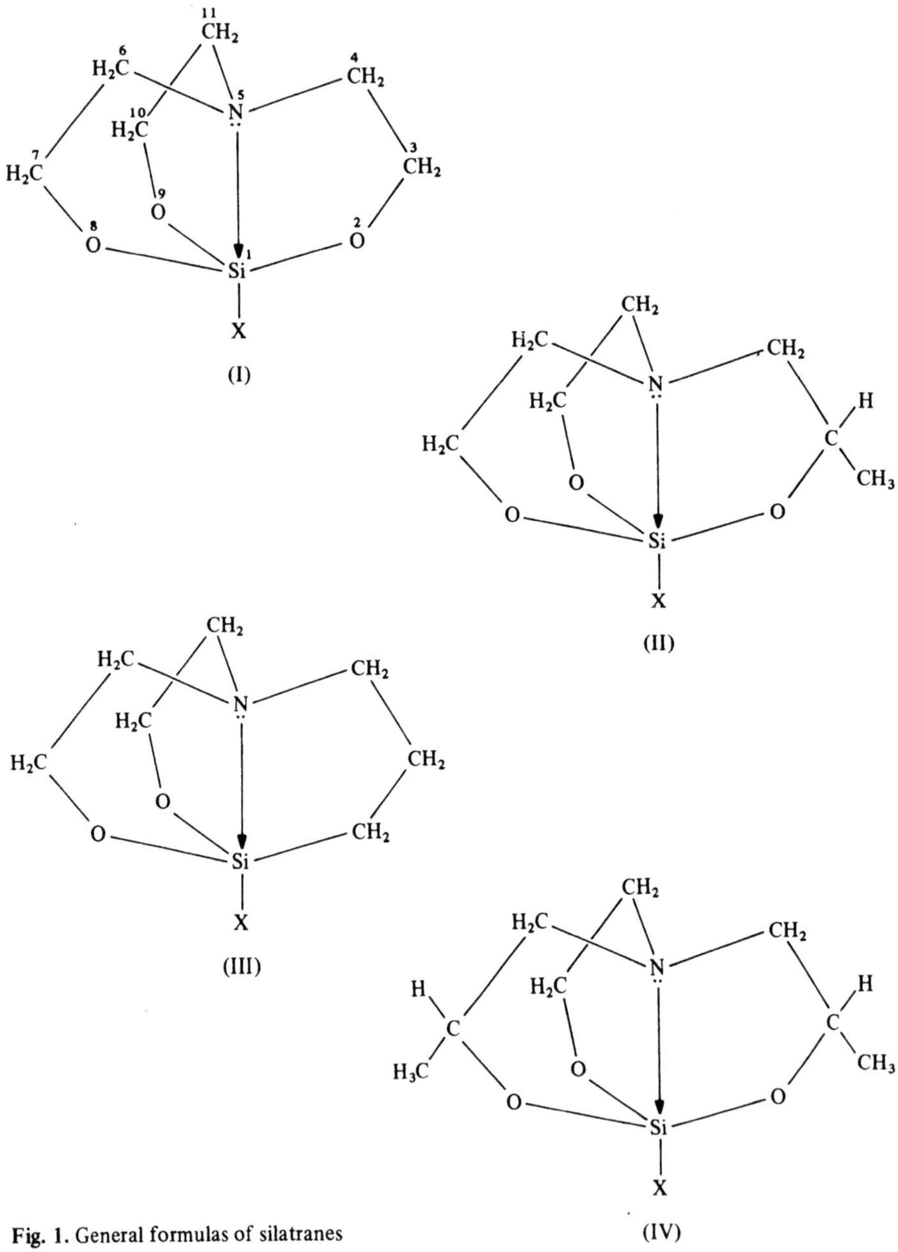

Fig. 1. General formulas of silatranes

During the following years investigations in the field of organosilicon compounds were concentrated on the development of new methods of synthesis of silatranes and on the study of their steric and electronic structure, various physical properties and, especially, their biological activity[2–15].

The discovery of the specific physiological activity of silatranes provoked an extensive search for new types of biologically active organosilicon compounds and raised the question of the creation of a new branch of silicon chemistry – bioorganosilicon chemistry[2].

Investigations in this field were carried out by a large team of scientists of different specialities such as biologists, microbiologists, pharmacologists, physiologists, physicians, agrochemists, etc. These investigations have shown that specific biological activity is displayed by many practically non-toxic or low toxicity silatranes. A number of them has proved very promising for applications in therapy, agriculture and industrial microbiology.

Some characteristic results on biological activity of silatranes and the possibility of their application will be considered in the following.

1 Toxicity

Today the toxic effect of silatranes has been thoroughly studied[1–3, 16–44a, b]. The toxicity of silatranes varies greatly and is mainly dependent on the nature of substituents at the silicon atom. Most toxic of all known silatranes are 1-arylsilatranes, $4\text{-}XC_6H_4Si(OCH_2CH_2)_3N$, where $X = CH_3$, Cl, H (Table 1). These compounds are almost twice as toxic as such well-known poisons as strychnine and hydrocyanic acid. They produce an intensive stimulation of the motor and respiratory centres[1–3, 30–34, 44] when administered at lower than lethal doses

Table 1. Toxicity of 1-arylsilatranes and their analogues (compare Fig. 1)

Compound	X	LD_{50}[mg/kg]	Compound	X	LD_{50}[mg/kg]
I	$4\text{-}CH_3C_6H_4$	0.15–0.20	I	$3\text{-}ClC_6H_4$	4.4
I	$4\text{-}ClC_6H_4$	0.22	III	C_6H_5	8.1
		0.9	I	$2\text{-}ClC_6H_4$	11.3
I	C_6H_5	0.3–0.5	IV	C_6H_5	8.1
		1.4[a]	I	$2\text{-}CH_3C_6H_4$	34
II	C_6H_5	0.5	I	$3\text{-}CH_3C_6H_4$	73
I	$5\text{-}CH_3C_4H_2S$	0.42	I	$\underline{CH_2(CH_2)_4CH}$	150
I	$5\text{-}BrC_4H_2S$	0.42	I	$C_6H_5CH_2$	1115
I	$5\text{-}ClC_4H_2S$	0.50			
I	C_4H_3S	1.70			
I	$4\text{-}FCH_2C_6H_4$	3.5			

[a] Per os (E. Bien, 1971).

Table 2. Oral toxicity of 1-(4'-chlorophenyl)silatrane
(C. B. Beiter, M. Schwarcz, G. Crabtree, 1971)

Species of warm blooded animals	LD_{50}[mg/kg]
Sparrows	0.2–0.4
White mice	0.9–2.0
Rats (Rattus norvegicus)	1–4
Mallard ducks	5–10
Pintail ducks	5–10
Monkey	14.0

(LD_{50} is 0.15–0.40 mg/kg for white mice). At low doses 1-arylsilatranes produce a local anaesthetic effect[39, 40].

The toxicity of 1-(4'-chlorophenyl)silatrane is highly selective (Table 2). This compound is most toxic for sparrows and mice and almost 100 times less toxic for monkeys[30–32].

The toxicity of 1-phenylsilatrane has been investigated on intraperitoneal administration to white mice[1]. When administrated at a dosage of only 0.20–0.25 mg/kg this compound induces strong motor stimulation typical for the effect of morphine, and shortness of breath. At slightly higher doses (0.35 mg/kg) the above manifestations are followed by tonic-clonic spasm. At doses of about 0.4 mg/kg this spasmodic effect often culminates in death. The average lethal dosage of 1-phenylsilatrane is 0.43 (0.36–0.50) mg/kg with $P = 0.05$.

The toxic effect of 1-phenylsilatrane on unanesthetized cats is similar to that described above. When administered intraperitonealiy at 0.30 mg/kg this preparation causes, after a short period of muscular twitching of the ears and lids, strong tonic-clonic convulsions and death within 8 minutes. However, in cats anaesthetized with urethane (1 g/kg), this toxic effect of 1-phenylsilatrane appears much weaker. Even intravenous administration at doses 10-fold greater than lethal does not lead to death. Given at doses of 0.30–0.40 mg/kg 1-phenylsilatrane induces only spasmodic twitching of skeletal muscles accompanied by respiratory stimulation.

The toxicity of 1-arylsilatranes is also reduced by barbiturates[33–37, 44]. Intravenous administration of 0.20 mg/kg of 1-phenylsilatrane to unanaesthesized rats induces motor stimulation and spasmodic twitching of skeletal muscles. The dosage of 0.30 mg/kg causes tonic-clonic spasm lasting several minutes. These doses do not give rise to mortality.

Frogs are highly resistant to 1-phenylsilatrane. Even doses as high as 30–40 mg/kg do not affect their behaviour.

The mechanism of muscular twitching and convulsions under the influence of 1-phenylsilatrane was studied on isolated rectal abdominal muscles of frogs. The compound used at 10^{-5} M concentration does not cause contraction of the muscle. No marked effect is observed after 10^{-5} M 1-phenylsilatrane acting for 2 minutes upon the contraction of the rectal abdominal muscle which had been induced by 10^{-7} M acetylcholine. Isolated gastrocnemius muscles of frogs bathed in 0.01%

1-phenylsilatrane in Ringer solution also show no contraction. Perfusion of hind limbs of the frog with the same solution also fails to elicit a response.

All these exclude the possibility of a direct effect of 1-phenylsilatrane on the peripheral neuromuscular apparatus.

In experiments with decerebrated cats the spasm and respiratory stimulation were observed immediately after administration of 1-phenylsilatrane. The preliminary pharmacological denervation of the sinocarotid regions with 0.5% novocaine solution (5 ml) did not affect the respiratory stimulation. Spinal cats displayed spasmodic muscular twitching after administration of 1-phenylsilatrane at higher doses.

Intravenous administration of this preparation to cats with completely destroyed central nervous system produces no effect even at doses as high as 1.5 mg/kg. This fact may indicate, first of all, the direct action on the central nervous system and the spinal cord. However, unlike other known compounds stimulating the spinal cord, 1-phenylsilatrane causes no similar effect on frogs. 1-Phenylsilatrane is thought to affect an enzymatic system of warm-blooded animals which is lacking in frogs or is not important for their central nervous system, since cholinesterase and acetylcholinesterase are not depressed by 1-phenylsilatrane[1, 2]. However American scientists think[36, 37] that the direct action of 1-phenylsilatrane which produces convulsive seizures and death of the animals is cardiac depression. The dead animals had injured ventricles that had caused the cardiovascular collapse. Blood coagulation increases in this case producing additional intravascular thromboses. These are the factors responsible for the death of the animals. Nevertheless the opinion prevails that 1-arylsilatranes destroy some of the nervous functions via the receptors of the spinal cord and the brain. This greatly stimulates motor activity[1, 2, 44), including that of the cardiac muscle.

The toxicity of 1-arylsilatranes is dependent not only on the nature of the substituent but also on the position of the latter in the aromatic ring. Thus, whereas para-methyl and para-chloroderivatives of 1-phenylsilatrane are highly toxic, their meta- and ortho-isomers show a low toxicity compared to them (Table 1). Introduction of the fluorine atom into the methyl group of 1-(4-methylphenyl)silatrane decreases its toxicity by one order of magnitude.

Hydration of the benzene ring in 1-phenylsilatrane leads to a sharp decrease of its toxicity (the LD_{50} value of 1-cyclohexylsilatrane is 150 mg/kg). The toxicity of 1-arylsilatranes drops even more on separation of the aromatic ring from the silicon atom by the methylene group (LD_{50} of 1-benzylsilatrane is 1115 mg/kg) (Table 1).

Introduction of methyl groups into the silatrane skeleton of 1-arylsilatranes also reduces considerably the toxicity (LD_{50} of 1-phenyl-3,7-dimethylsilatrane is two orders of magnitude lower than that of 1-benzylsilatrane) (Table 1).

According to a private communication of D. Bennett, USA, introduction of methyl groups into the silatrane skeleton of 1-phenylsilatrane reduces significantly its toxicity (Table 1, 3).

The toxic effect of all the above compounds begins to be manifested within 1–5 min. In lethal cases all animals die within 15 minutes. Bennett thinks that the toxic effect of 1-phenylsilatrane and its derivatives is due to central cholinergic stimulation.

Table 3. Comparative toxicity of 1-phenylsilatrane and its C-methylsubstituents (D. Bennett, 1974)

Compound	Convulsions [%]	Mortality[a] [%]
$C_6H_5Si[OCH_2CH_2]_3N$	100	100
$C_6H_5Si[(OCH_2CH_2)_2/OCH_2C(CH_3)_2]N$	100	20
$C_6H_5Si[(OCH_2CH_2)_2/OCH(CH_3)CH_2]N$	60	0

[a] On intraperitoneal administration at a 1 mg/kg dose (methacell suspension) to Badberg male rats.

The toxicity of 1-phenylsilatrane also is greatly reduced by introduction into position 3 of the CH_2Cl, $CH_2 = CH$ and, especially, C_6H_5 groups (LD_{50} = 3.4, 3.0, ~2000 mg/kg, respectively)[37a]. Si-substituted 3-phenylsilatranes are low toxic ($LD_{50} > 100$ mg/kg). The CH_2Cl goup in position 3 greatly increases the toxicity of 1-ethyl- and 1-vinylsilatranes (LD_{50} is approximately 100 and 30 mg/kg, respectively).

In 1971 M & T Chemicals (USA) began marketing 1-para-chlorophenylsilatrane as a new rodenticide under the trade mark RS-150[30, 31]. This first organosilicon pesticide had an advantage over other known toxicants. This highly toxic compound is rapidly inactivated in poisoned rodents, so their corpses are not harmful to other animals. Furthermore, RS-150 penetrates very poorly through the skin (its cutaneous toxicity is 3000 mg/kg for rats)[31]. Finally, rats do not become resistant to 1-(4′-chlorophenyl)silatrane as was the case with other 1-arylsilatranes. Practically complete lack of smell is another important advantage of this preparation and other 1-arylsilatranes.

1-(4′-Chorophenyl)silatrane destroys nervous functions in the central nervous system of vertebrates, primarily in the brain and possibly in the brain stem[44]. Phenobarbital alleviates the convulsions in mice caused by compounds of this type.

1-(2′-Thienyl)silatranes are similar to arylsilatranes in their toxicity (Table 1)[43a]. Introduction of a chlorine or bromine atom or a methyl group into the 5-thiophene ring causes no significant changes in toxicity. The blood pressure of urethane anaesthetized cats is not influenced by these compounds even at a 2 mg/kg dose.

Intraperitoneal injections of 1-(2′-thienyl)silatrane into rats causes tremor culminated in convulsions within some minutes and in death if lethal doses are given[44a, 44b]. The convulsions are effectively antagonized by phenobarbital (100 mg/kg) which also reduces the toxicity of the preparation. The toxic effect of 1-(2′-thienyl)silatrane also is reduced by diazepam (5 mg/kg) but to a less extent. Atropine and 3-quinoclidinyl benzylate do not affect the silatrane toxicity. The symptoms of poisoning indicate that 1-(2′-thienyl)silatrane activates the central nervous system. Both convulsive and subconvulsive doses of this preparation elevate cyclic GMP concentrations in the cerebellum. At the same time, no significant changes in the cyclic GMP levels are observed in cerebral cortex, brain stem and subcortial tissue. The cyclic AMP concentrations are not changed in any of the studied brain areas. Phenobarbital and diazepam reduce the increase of cyclic GMP. The effect of 1-(2′-thienyl)silatrane on the cyclic nucleotide levels in the rat brain

Table 4. Toxicity of 1-organoxysilatranes $ROSi(OCH_2CH_2)_3N$ and hydrolysis rate constants [J. E. Casida et al. (1976), M. G. Voronkov (1977)]

R	LD_{50}[mg/kg]	$k_2 \cdot 10^{-6}$ $\frac{mole}{l \cdot sec}$
C_2H_5	2800[a]	
	3000	1.48 ± 0.01
C_3H_7	3000	1.60 ± 0.01
$(CH_3)_2CH$	3000	2.04 ± 0.01
$C_6H_5CH_2$	2250	2.28 ± 0.02
CH_3	2100	
C_2H_5[b]	2000	
$(CH_3)_3CCH_2$	1700	1.88 ± 0.01
4-$(CH_3)_3CC_6H_4$	1400	
2-$CH_3C_6H_4$	1200	1.52 ± 0.01
$C_{10}H_7$	850	
4-$CH_3C_6H_4$	710	1.44 ± 0.01
3-$CH_3C_6H_4$	708	1.40 ± 0.01
4-$O_2NC_6H_4$	700	
2-ClC_6H_4	600	
4-ClC_6H_4	565	1.56 ± 0.02
4-$CH_3OC_6H_4$	345	1.48 ± 0.01
$(CH_3)_3C$	230	0.48 ± 0.02
C_6H_5	200	1.32 ± 0.01

[a] Per os (E. Bien, 1971).
[b] 1-Ethoxy-3-methylsilatrane.

is similar to that obtained with γ-aminobutyric acid (GABA) receptor blockers and bicyclic phosphorus erters which are suggested to act as GABA antagonists. However, the symptoms induced by injections of the latter and 1-(2′-thienyl)silatrane are not identical.

1-Alkoxysilatranes have low toxicity and most of them are practically non-toxic (Table 4).

Thus, after intraperitoneal administration, 1-ethoxysilatrane produces no signs of poisoning even at doses higher than 1500 mg/kg[1] ($LD_{50} > 3000$ mg/kg). Stable to hydrolysis, 1-tertbutoxysilatrane is the most toxic ($LD_{50} = 230$ mg/kg) of all 1-alkoxysilatranes. Readily hydrolyzable 1-methoxysilatrane is less toxic than 1-tert-butoxysilatrane but more toxic than 1-ethoxysilatrane and its nearest homologues (Table 4)[43]. This slightly higher toxicity of 1-methoxysilatrane seems to be due to the influence of the methyl alcohol formed by hydrolysis. C-Substituted 1-ethoxy-silatranes are more toxic than 1-ethoxysilatrane.

1-Aroxysilatranes are more toxic than 1-alkoxysilatranes; their LD_{50} values range within 1400–200 mg/kg (Table 4). Substituents introduced into the para-position of the benzene ring of 1-phenoxysilatrane cause an increase in toxicity as follows $(CH_3)_3C(1400) < CH_3(710) < NO_2(700) < Cl(565) < CH_3O(345)$

The toxicity of 1-aroxysilatranes is definitely related to their stability in the animal organism. The data of Table 4 show that the toxicity decreases with the hydrolysis rate[45]. This implies that the poisoning effect is produced by 1-aroxy-

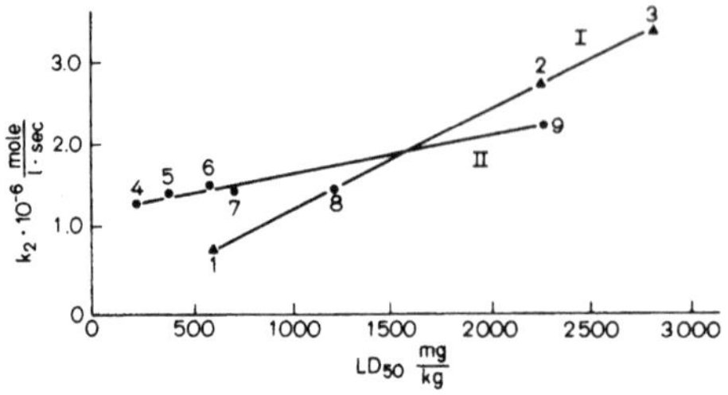

Fig. 2. k_2/LD_{50} Dependence for silatranes, $XSi(OCH_2CH_2)_3N$
I – 1-(chloroalkyl)silatrane,
$k_2 = 153.2 \cdot 10^{-3} + 1.2 \cdot 10^{-3}\, LD_{50}$;
$r = 0.999$, $S_0 = 0.4 \cdot 10^{-6}$;
II – 1-aroxysilatranes,
$k_2 = 1206.8 \cdot 10^{-3} + 0.43 \cdot 10^{-3}\, LD_{50}$;
$r = 0.918$, $S_0 = 0.10 \cdot 10^{-6}$

silatranes rather than by their hydrolysis products. The same is indicated by a satisfactory linear relationship between the LD_{50} and the lgk value
$lgk = 0.50 - 1.05 \cdot 10^{-4}\, LD_{50}$ ($Z = 0.91$, $S = 0.02$) (Fig. 2).

Withdrawal of the aromatic substituent in 1-aroxysilatranes from the oxygen atom in a methylene group leads to a sufficient decrease in toxicity (LD_{50} of 1-benzyloxysilatrane is 2250 mg/kg). Whereas administration of 150 mg/kg of 1-phenoxysilatrane to white mice produces tonic-clonic convulsions, 1-benzyloxysilatranes administered at higher doses only cause slight hyperemia.

1-Alkyl-, 1-vinyl-, and 1-ethynylsilatranes are practically non-toxic (Table 5) (the LD_{50} value exceeds 3000 mg/kg). Administration of these compounds at 3000 mg/kg to white mice does not give rise to mortality. Only injection of 1-ethynylsilatrane produces slight excitation.

Substitution of the acetylenic hydrogen atom in 1-ethynylsilatrane by a phenyl group slightly increases the toxicity. The presence of bi- or tricyclic hydrocarbon radicals at the silicon atom leads to a greater increase in toxicity (the LD_{50} value for compounds (30) and (41) is 850 and 80 mg/kg, respectively).

Introduction of functional groups into a hydrogen radical in 1-organylsilatranes produces great variation in toxicity. Thus the presence of acryloxy- (7), methacryloxy- (8), aroyloxy- (14–16), and aroxyacetoxy- (17 and 21) groups in the methyl group of 1-methylsilatrane increases negligibly the toxicity (LD_{50} is 2500–3000 mg/kg) or does not change it at all.

Nitrogen-containing 1-alkylsilatrane derivatives (9, 23, 24 and 25) have low toxicity (LD_{50} = 1300 – 3000 mg/kg). However, quaternization of the nitrogen atom in 1-aminoalkylsilatranes (32, 33) leads to a sharp increase in toxicity (LD_{50} is 540 and 240 mg/kg, respectively). 1-(3′-Aminopropyl)silatrane (36) is even more toxic (LD_{50} = 190 mg/kg).

Sulphur derivatives of 1-alkylsilatranes are commonly toxic[46, 47]. Thus, 1-mercaptoalkylsilatranes[26, 27] are almost 2 orders more toxic than the initial mercaptoalkyltrialkoxysilanes[48].

1-(Benzylthiomethyl)silatrane (40) and 1-ethylthiomethylsilatrane (46) are fairly toxic too (LD_{50} = 95 and 16 mg/kg, respectively). At the same time, intraperitoneally administrated 1-n-propylthiomethylsilatrane (22) and bis(silatranylmethyl)sulphide, $[N(CH_2CH_2O)_3SiCH_2)]_2S$ are not toxic (LD_{50} = 4000 mg/kg).

1-(Thiocyanatoalkyl)silatranes, $NCS(CH_2)_nSi(OCH_2CH_2)_3N$, (50 and 58) are highly toxic[49]. For the compounds having n = 1, 3 the LD_{50} value is 2.0 and 0.2 mg/kg, respectively.

Table 5. Average lethal doses (LD_{50}) of 1-substituted silatranes $XSi(OCH_2CH_2)_3N$

No.	X	LD_{50}[mg/kg]	No.	X	LD_{50}[mg/kg]
1	CH_3	3000	31	Cl_2CH	600
2	CH_3CH_2	3000	32	$[CH_3(C_2H_5)_2NCH_2]^+I^-$	540
3	$(CH_3)_2CH$	3000	33	$[\overline{CH_2(CH_2)_3N}(CH_3)CH_2]^+I^-$	240
4	$CH_2{=}CH$	3000	34	$FCH_2CH_2CH_2$	223
5	$CH{\equiv}C$	3000	35	$C_6H_5(CH_3)CH$	200
6	$C_6H_5C{\equiv}C$	3000	36	$H_2NCH_2CH_2CH_2$	190
7	$CH_2{=}CHCOOCH_2$	3000	37	$C_6H_5N(CH_3)CH_2$	170
8	$CH_2{=}C(CH_3)COOCH_2$	3000	38	$\overline{CH_2(CH_2)_4CH}$	150
9	$O(CH_2CH_2)_2NCH_2$	3000	39	H	100
10	$(C_2H_5O)_2P(O)CH_2$	3000	40	$C_6H_5CH_2SCH_2$	95
11	CH_3CHCl	3000	41	C_7H_{11} [d]	80
12	$C_6H_5CH{=}CH$	3000	42	$HSCH_2CH_2$	55
13	$C_6H_5CH_2CH_2$	3000	43	$HSCH_2$	53
14	$4\text{-}ClC_6H_4COOCH_2$	3000	44	$ICH_2CH_2CH_2$	29
15	$4\text{-}BrC_6H_4COOCH_2$	3000	45	$4\text{-}CH_3OC_6H_4$	17
16	$4\text{-}CH_3C_6H_4COOCH_2$	2900	46	$C_2H_5SCH_2$	16
17	$2\text{-}CH_3C_6H_4OCH_2COOCH_2$	2800	47	$4\text{-}(C_2H_5)_2NC_6H_4$	10
18	$ClCH_2$	2800	48	$BrCH_2CH_2CH_2$	5.0
19	$ClCH_2CH_2CH_2$	2240	49	$3\text{-}ClC_6H_4$	4.4
20	CH_3OCH_2	2100	50	$NCSCH_2$	2.0
21	$4\text{-}ClC_6H_4OCH_2COOCH_2$	2100	51	$4\text{-}ClC_6H_4$	1.7
22	$CH_3CH_2CH_2SCH_2$	2000	52	$2\text{-}C_4H_3S$ [e]	0.76
23	$\overline{CH_2(CH_2)_4N}CH_2$	1800	53	$5\text{-}ClC_4H_2S$	0.50
24	$3.5\text{-}(CH_3)_2C_3HN_2CH_2$ [a]	1500	54	$5\text{-}BrC_4H_2S$	0.42
25	$NCCH_2CH_2$	1340	55	$5\text{-}CH_3C_4H_2S$	0.42
26	$C_6H_5CH_2$	1115	56	$2.4\text{-}Cl_2C_6H_3$	0.42
27	$C_4H_3OSi(CH_3)_2CH_2CH_2$ [b]	1000	57	C_6H_5	0.33
28	$BrCH_2$	916	58	$NCS(CH_2)_3$	0.20
29	ICH_2	890	59	$4\text{-}CH_3C_6H_4$	0.15
30	$C_{10}H_{15}CH_2CH_2$ [c]	850			

a 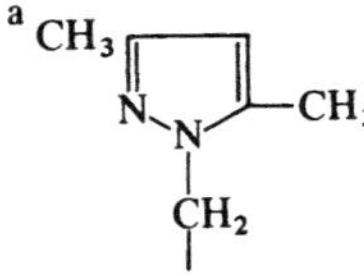

b 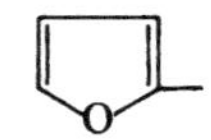

c 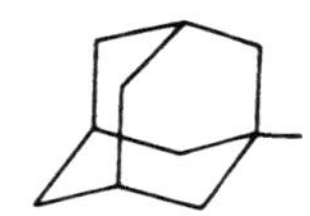

d 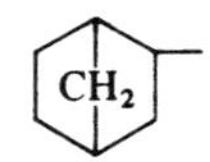

e 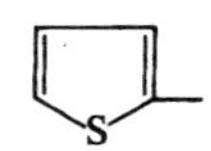

Introduction of a $(C_2H_5O)_2P(O)$ group (10) into a molecule of 1-alkylsilatranes does not change their toxicity whereas triorganyl(silatranylmethyl)phosphonium iodides are highly toxic for white mice. For example, the LD_{50} value of triphenyl (silatranylmethyl)phosphonium iodide $[N(CH_2CH_2O)_3SiCH_2P(C_6H_5)_3]^+I^-$ is 2.2 mg/kg.

Application of this compound immediately involves general excitability, slow respiration, convulsions and contraction of the abdominal muscles culminating in coma 7–10 minutes after treatment.

Introduction of a chlorine atom into the 1- or 3-position of the alkyl radical in 1-alkylsilatranes(18, 19) decreases to some extent the average LD_{50} value 2200–2800 mg/kg). However almost all 1-chloroalkylsilatranes prove to have low toxicity. Replacement of one hydrogen atom in the methyl radical of 1-methylsilatrane by a bromine or iodine atom (28, 29) as well as the replacement by two halogen atoms (31) increases markedly the toxicity of 1-halogenalkylsilatranes. In this case the animals show depression, rapid breathing, tremor. Administration of 1-bromomethyl- (28) and 1-iodomethylsilatrane (29) causes hyperaemia, tremor and depression.

Table 6. Toxicity of fluoro-containing silatranes

	Compound	LD_{50}[mg/kg]
1	$C_6F_{13}CH_2CHISi(OCH_2CH_2)_3N$	3000
2	$ClCH_2Si{-}$ OCH_2CH_2 / OCH_2CH_2 / $OCH(CF_3)CH_2$ $-N$	893
3	$ClCH_2Si[OCH(CF_3)CH_2]_3N$	268
4	$CF_3(CH_2)_2Si{-}$ OCH_2CH_2 / OCH_2CH_2 / $OCH(CH_3)CH_2$ $-N$	260
5	$CF_3(CH_2)_2Si{-}$ OCH_2CH_2 / $OCH(CH_3)CH_2$ / $OCH(CH_3)CH_2$ $-N$	250
6	$CF_3(CH_2)_2Si[OCH(CH_3)CH_2]_3N$	240
7	$FCH_2(CH_2)_2Si(OCH_2CH_2)_3N$	220
8	$CF_3CH_2CHISi(OCH_2CH_2)_3N$	120
9	$FCH_2(CH_2)_2Si[OCH(CH_3)CH_2]_3N$	90
10	$CF_3(CH_2)_2Si{-}$ OCH_2CH_2 / OCH_2CH_2 / $OCH(CF_3)CH_2$ $-N$	54
11	$CF_3(CH_2)_2Si(OCH_2CH_2)_3N$	5

The toxicity of 1-(3′-halopropyl)silatranes (34, 44, 48) is dependent on the nature of the halogen atom and increases in the following order

Cl (2250) < F (220) < J (29) < Br (5)

(figures in brackets show the LD_{50} values in mg/kg).

Introduction of FCH_2 and F_3C groups into the side chain of 1-alkylsilatranes increases the toxicity more effectively than introduction of trifluoromethyl groups into the silatrane skeleton (Table 6).

The clinical signs after administration of compounds containing fluorine are almost identical. Death occurs more rapidly with higher toxicity of the compound. Thus, administration of 1-(3′-trifluoropropyl)silatrane and 1-(3′-fluoropropyl)-3,7,10-trimethylsilatrane at the highest dosage causes immediate death while in the case of 1-(3′-trifluoropropyl)-3,7,10-trimethylsilatrane death occurs within 10–15 minutes[42, 50].

The immediate cause of death seems to be the result of a paralytic action on the central nervous system, particularly on the respiratory centre. Typical toxic signs of poisoning by fluorine-containing silatranes are hyperexcitation, muscle twitching, spasm of abdominal muscles, tremor, salivation.

Sometimes administration of 1-(3′-trifluoropropyl)-3,7,10-trimethylsilatrane induces epistaxis.

Polyfluoroorganylsilatranes containing an iodine atom at the α-position to the silicon atom are less toxic than 1-(3′-trifluoropropyl)silatrane (11) (Table 6). The LD_{50} value for $CF_3CH_2CHISi(OCH_2CH_2)_3N$ is 120 mg/kg and that for $C_3F_7CH_2CHISi(OCH_2CH_2)_3N$ is 180 mg/kg. At the same time, 1-(2′-perfluorohexyl-1′-iodoethyl)silatrane (1) is practically non-toxic (LD_{50} = 3000 mg/kg).

The toxicity of certain types of carbofunctional substituted 1-organylsilatranes is related, to a certain degree, to the hydrolysis rate. Thus, the toxic effect of 1-chloroalkylsilatranes as well as that of 1-aroxysilatranes is linearly related to the hydrolysis rate constant (Fig. 2).

The toxicity of silatranes is considerably influenced by the nature of the solvent. Thus, the toxicity of silatrane solutions in dimethylsulphoxide is considerably higher than in water and tween. On skin application 1-arylsilatranes in DMSO solutions are fairly toxic as well, but not toxic in the solid state and in aqueous solutions.

The mechanism of the toxic effect of silatranes is not clear yet. The data of Tables 1–6 show that toxicity depends, first of all, on the substituent at the silicon atom.

Some attempts have been made to explain the high toxicity of 1-arylsilatranes by the presence of the transannular Si-N bond and pentacoordinate silicon in their molecules[1, 2, 34].

During the last years high toxicity was found for a series of bicyclic systems similar to silatranes[44, 51] such as phosphatranes, azaadamantanes and trioxabicyclo (2.2.2)octanes (Table 7). This shows that the high toxicity of silatranes[1] results to a considerable extent from stereochemical factors. It is connected, first of all, with

1 This is true for many other biological properties.

Table 7. Toxicity of polyhedral systems of some silatranes, phosphatranes, heteroadamantanes and bicyclo[2.2.2]octanes[a]

Compound	LD_{50}[mg/kg]
$CH_3Si(OCH_2CH_2)_3N$	3000
$ClCH_2Si(OCH_2CH_2)_3N$	1500
$BrCH_2Si(OCH_2CH_2)_3N$	916
$C_6H_5Si(OCH_2CH_2)_3N$	0.3
$(S)P(OCH_2CH_2)_3N$	260
$(O)P(OCH_2)_3P(O)$	189
$[(C_6H_5)_3CP(OCH_2CH_2)_3N]BF_4$	28
$(O)P(OCH_2)_3CC_6H_5$	1.5
$(O)P(OCH_2)_3CC_3H_7$-n	0.39
$(S)P(OCH_2)_3CC_3H_7$-i	0.26
$P(OCH_2)_3CC_3H_7$-i	0.22
$(O)P(OCH_2)_3CC_3H_7$-i	0.18
$ClCH_2C(OCH_2)_3CCH_3$	526
$CH_3C(OCH_2)_3CCH_3$	500
$CH_3C(OCH_2)_3CCH_2Br$	390
$HC(OCH_2)_3CC_3H_7$-i	30
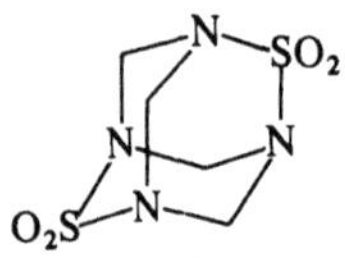	0.24
Strychnin[b]	5–15

[a] J. E. Casida, M. Eto, A. D. Moscioni, J. L. Engel, D. S. Milbrath, J. G. Verkade, (1976)[44]

[b] lethal dose for human (Z. Franke, 1973)[52]

the spheric arrangement of the silatrane cycle and the presence of the "bridging" nitrogen atom. However, in silatranes the nature of the silicon atom has a considerably greater influence on the toxicity than in other polyhedral structures. For example, for 1-oxophospha-2,6,7-trioxa-4-organylbicyclo(2.2.2)octanes, $(O)P(OCH_2)_3CR$, the compounds with R = alkyl are even more toxic than their analogues with R = aryl[44]. At the same time, 1-alkylsilatranes are practically non-toxic, unlike 1-arylsilatranes.

The toxic effect and other biological properties of many silatranes are thus associated with their structural peculiarities and are mainly determined, in the author's opinion, by the following factors

1. Polyhedral structure of the silatrane skeleton and anomalous valency state of the silicon and nitrogen atoms involved.
2. High dipole moment of the molecule.
3. The nature of the substituent bound to the silatrane skeleton.
4. Rate of cleavage of the most reactive bonds in the molecule both within and outside the silatrane skeleton.

2 The Effect on Enzyme Activity and Some Biochemical Reactions

Data obtained on the effect of 1-(chloromethyl)silatrane on the enzyme activity and the nucleic acid and protein contents of liver and spleen of mature rats, are listed in Tables 8–10.

Table 8. The effect of 1-(chloromethyl)silatrane on some biochemical processes

Enzymatic process or biochemical reaction	Deviation from control values [%] Concentration [mol/l] 10^{-3}	10^{-5}	10^{-7}	Conclusion
Phosphodiesterase of rat brain	–5.1	–6.2	–6.2	No effect
Monoamine oxidase of mitochondria of rat liver	–2.7	–11	–	No effect
Acetylcholinesterase of rat brain	–50	0	0	Inhibitor
Butyrylcholinesterase of blood serum	0	0	0	No effect
Total adenosine triphosphatase activity of rat liver	+80	+62	+40	Activator
Linking with SH-groups of glutathione	+5	0	0	No reaction

Table 9. The effect of 1-(chloromethyl)silatrane on the protein, RNA and DNA contents of liver and spleen of rats, administered daily at 100 mg/kg for 3 days

Organ	Weight [% of control]	Content [% of control]		
		Protein	RNA	DNA
Liver	108	112	128	101
Spleen	110	122	123	110

Table 10. The effect of 1-(chloromethyl)silatrane on the DNA biosynthesis in regenerating liver of rats

Time of administration [hours after hepatectomy]	H^3-thymidine $[\frac{\text{pulse/min}}{\text{mg DNA}} 10^{-1}]$	Stimulation of DNA synthesis [%]	$t_d = \frac{M_d}{m_d}$
3	845.2 ± 105.0	29.8	2.22
8	1745.7 ± 125.0	66.7	9.50
17	1191.0 ± 131.0	51.5	4.77
19	1473.0 ± 83.2	54.5	10.85
Control	560.6 ± 25.5	0	0

Table 8 shows that at concentrations of 10^{-7} – 10^{-8} M 1-(chloromethyl)silatrane does not affect phosphodiesterase of the rat brain and monoamine oxidase of the rat liver. At the same time, at 10^{-4} M concentration the preparation weakly inhibits the acetylcholinesterase of the rat brain. Therefore, 1-(chloromethyl)silatrane may be expected to produce a gentle stimulatory effect on the processes in the central and peripheral nervous system which are mediated by acetylcholine. 1-(Chloromethyl)silatrane activates mildly the preparation of summarized ATP values of the rat liver. No reaction is observed with SH-groups of glutathione.

Table 9 shows the effect of 1-(chloromethyl)silatrane on the weight of liver and spleen, as well as the nucleic acid and protein contents of these organs. The data indicate that 1-(chloromethyl)silatrane does not influence the DNA content of the liver but it increases slightly (by 10%) that of the spleen. The RNA content of the liver and spleen increases considerably (by 28% and 23%, respectively). Simultaneously, the protein content of both organs increases by 12% and 22%. This shows that 1-(chloromethyl)silatranes stimulates the biosynthesis of RNA and protein in the organs studied. The stimulation effect of 1-(chloromethyl)silatrane on the biosynthesis is better manifested in spleen than in liver.

Normal liver cells are known to perform only functionally essential biosynthesis. The rate of mitoses of these cells is 0.8%. After partial hepatectomy (resection of 2/3 of liver) a sharp intensification of biosynthesis with accelerating cell division takes place.

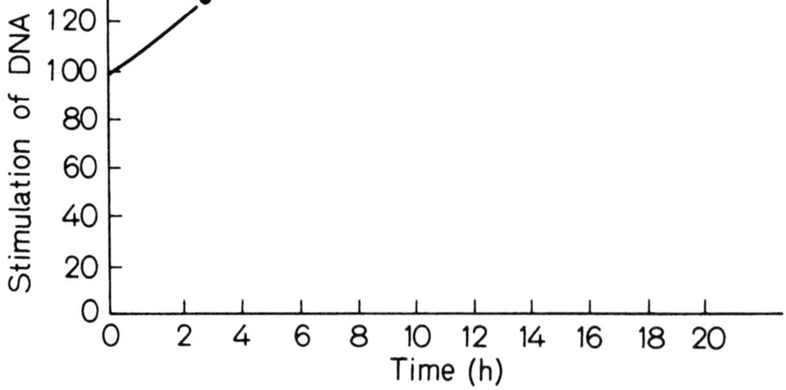

Fig. 3. Influence of 1-(chloromethyl)silatrane (100 mg/kg per day) on the DNA biosynthesis in the regenerating liver of rats

The stimulatory effect of 1-(chloromethyl)silatrane on the DNA synthesis in the cells of the regenerating liver is shown in Table 10 and Fig. 3. In a similar way (according to incorporation of ^{3}H-thymidine, ^{14}C-orotic acid and ^{3}H-leucine) it has been found that 1-(chloromethyl)silatrane intensifies the DNA, RNA and protein biosynthesis in other developing cells (by 20–60%). This has shown the importance of further investigation of this preparation as a stimulator of cell division and biosynthesis of nucleic acids and proteins.

The influence of 1-ethoxysilatrane on regenerating liver has been studied on male white mice after resection of 2/3 of the liver[53)]. The animals were treated daily with a therapeutic dose (100 mg/kg) of 1-ethoxysilatrane, the control animals only with physiological saline solution.

The regeneration process was checked by the growth of liver weight. 1-Ethoxysilatrane is found to intensify regeneration of liver substance; the weight gain was 11–12% 2–4 days after the resection. 1-Ethoxysilatrane produces favourable effects not only on the liver, but on the whole organism of the animal.

Further studies have shown the influence of 1-ethoxysilatrane on the activity and the size of the regenerating liver cells. The cell size of the treated animals is 32% larger than in controls[54)]. The total amount of mitotic phases in the treated animals increases by 73%.

The index of average mitotic phases (without previous prophase and telophase) increases too. This indicates that 1-ethoxysilatrane accelerates the initial and final steps of mitosis.

The above data suggest that silicon stimulates the biosynthesis of collagen as a coupling or, possibly, matrix agent[13, 15)]. In fact, further experiments showed that being excellent donors of essential silicon, some silatranes produced a marked stimulating effect on not only the total protein synthesis but on the synthesis of collagen as well[66)].

The effect of Si-substituted silatranes, $XSi(OCH_2CH_2)_3N$ ($X = CH_3, C_2H_5O, ClCH_2$), on the biosynthesis of collagen was studied in vitro using cartilagenous tissue of chick embryos. The dynamics of the process was determined by the formation of peptide-bound ^{14}C-oxyproline.

The most pronounced stimulation of the biosynthesis of collagen was observed in concentration of $3 \cdot 10^{-3}$ mole/l within 180 min of incubation (^{14}C-proline was added to the incubation medium 60 min after the beginning of incubation).

Of the three silatranes studied, 1-methylsilatrane produces the strongest effect on the total protein synthesis in the cartilagenous tissue of chick embryos.

The silatranes studied produce a clear stimulating effect on the formation of oxyproline in intact bones of chick embryos. However, these silatranes do not affect much the activity of collagen propyl-hydroxylase catalyzing the oxyproline formation.

This indicates that the stimulation of ^{14}C-oxyproline formation by silatranes is associated with the influence of the protein-synthesizing apparatus of the cell rather than with the activity of collagen prolyl-hydroxylase.

3 The Effect on Blood

The effect of silatranes of the type $XSi(OCH_2CH_2)_3N$ where $X = CH_3, C_2H_5$, CH_3CH_2O, $(CH_3)_2CHO$, $(CH_3)_3CO$ on blood has been studied in vitro[55]. The compounds were injected into native blood at 2.5, 5.0, 10.0, 20.0, 25.0 and 30.0 mg/ml doses. Blood coagulation was checked on glass and paraffin. In control tests physiological salt solution was used. All six silicon-substituted silatranes studied make coagulation slower both on glass and paraffin. The activity of these compounds is dependent on X and increases in the following order

$$(CH_3)_3CO < (CH_3)_2CHO < C_2H_5 < CH_3 < CH_3O < CH_3CH_2O.$$

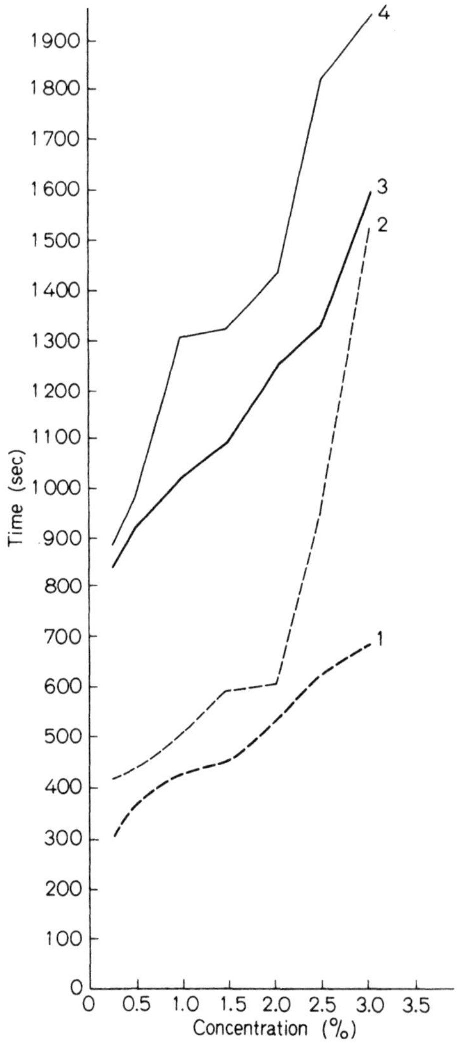

Fig. 4. Influence of 1-ethoxysilatrane on blood coagulation in vitro. Glass: start (*1*), finish (*3*). Paraffin: start (*2*), finish (*4*)

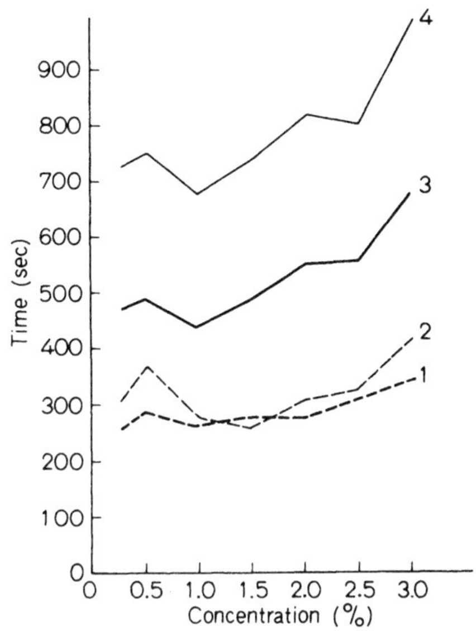

Fig. 5. Influence of 1-isopropoxysilatrane on blood coagulation in vitro.
Glass: start (*1*), finish (*3*). Paraffin: start (*2*), finish (*4*)

The influence of 1-ethoxysilatrane and 1-isopropoxysilatrane on blood coagulation is shown in Figs. 4 and 5 respectively. It is seen that both 1-alkoxysilatranes are more effective in the end than in the beginning of coagulation. 1-Isopropoxysilatrane at a dose of 5–20 mg/ml causes no coagulation.

1-Ethoxysilatrane displays a considerably larger anticoagulant effect than all other silatranes investigated[56)]. The data on the effect of some silatranes on blood coagulation are represented in Table 11. They show that the silatranes examined increase thrombin time appreciably only when added to plasma at high doses (1000–5000 γ/ml). Thromboplastin time is practically unaffected by all silatranes studied, except 1-ethoxysilatrane which raises this factor more than 2-fold at 5000 γ/ml dosage. 1-(Chloromethyl)-3,7,10-trimethylsilatrane and 1-ethoxysilatrane considerably increase plasma recalcification time.

Table 11. The influence of silatranes on haemocoagulation indices of the blood plasma

Compound	Dosage [γ/ml]	Recalcification time [sec]	Thrombin time [sec]	Thromboplastin time [sec]
1-(Chloromethyl)silatrane	1000	182	117	32
1-(Chloromethyl)-3,7,10-trimethylsilatrane	1000	222	51	22
1-Ethoxysilatrane	5000	228	168	48
1-Ethoxy-3,7,10-trimethylsilatrane	5000	145	44	25
Control (Physiologic solution)	0	100	20	23

$[(C_6H_5)_3PCH_2Si(OCH_2CH_2)_3N]^+I^-$produces a stabilizing effect on erythrocyte membranes at concentrations of 1 γ/ml. The zone of the minimum osmotic resistance of erythrocytes is 0.46% of the NaCl concentration and the zone of the maximum osmotic resistance is 0.38%, thus exceeding the control values[57].

1-(0,0-Diethylphosphonemethyl)-3,7,10-trimethylsilatrane and its analogs may prove very useful in treatment of anaemia accompanied with intensive erythrocytolysis, in blood conservation and haemotransfusion therapy to prevent destruction of the erythrocyte membranes.

The effect of this silatrane on the acid resistance of erythrocytes has been examined using a model anaemia induced by a heavy metal salt[58]. For this purpose 2 ml of 0.001% thallium nitrate and 2 ml of 0.1% or 0.01% 1-(0,0-diethylphosphonemethyl)-3,7,10-trimethylsilatrane solution were added to the blood of rats. Then the blood was incubated for 1 hour at 37 °C.

Administration of this silatrane, especially at 0.01% concentration, to the erythrocytes incubated with thallium nitrate increases their resistance to haemolysis. The erythrograms shift to the right and the maximum erythrocytolysis decreases to 25% whereas the resistance time amounts to 600 ± 30 sec (Fig. 6).

On administration of 0.1 and 0.01% of 1-(0,0-diethylphosphonemethyl)-3,7,10-trimethylsilatrane in physiologic saline solution to the blood of animals poisoned with lead nitrate a sharp increase in acid resistance of erythrocytes was oberved (Fig. 7) along with erythrogram shift to the right and a prolonged prelytic time (to 240 ± 60 sec). Sometimes the maximum erythrocytolysis occurs only after 300 ± 60 sec. Haemolysis is completed in 720 ± 30 sec.

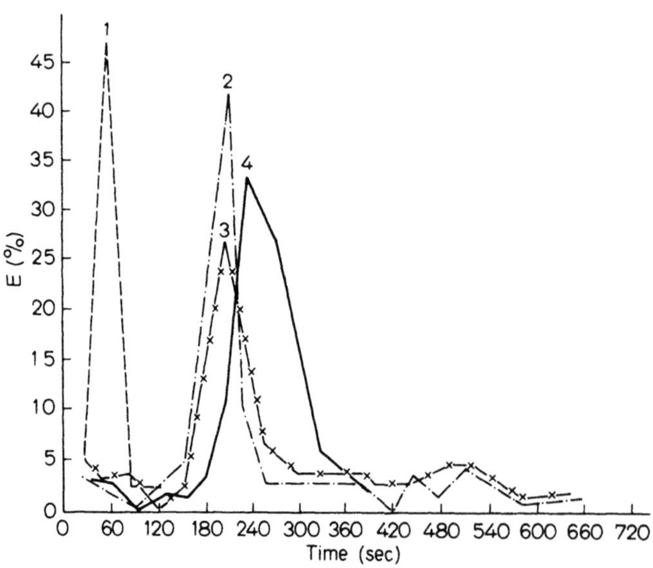

Fig. 6. Influence of $(C_2H_5O)_2P(O)CH_2Si[OCH(CH_3)CH_2]_3N$ (I) on acidic resistance of erythrocytes. Acidic erythrograms of blood after addition of: *1*) 0.001% $Tl(NO_3)_3$ solution, *2*) 0.001% $Tl(NO_3)_3$ solution and 0.1% (I), *3*) 0.001% $Tl(NO_3)_3$ solution and 0.01% (I), *4*) control

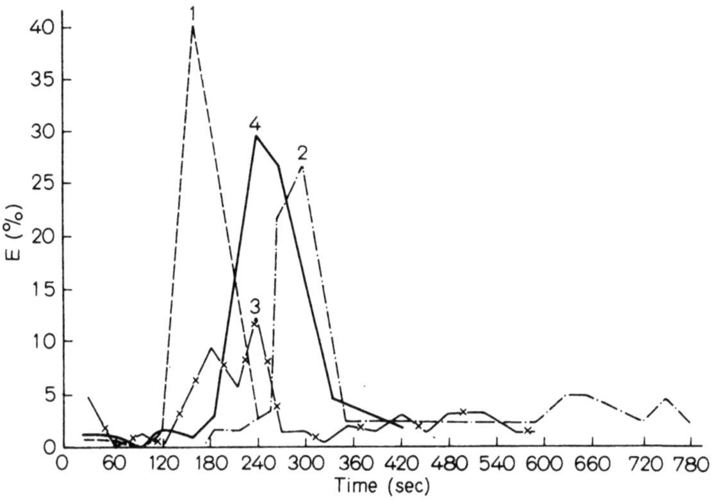

Fig. 7. Influence of $(C_2H_5O)_2P(O)CH_2Si[OCH(CH_3)CH_2]_3N$ (I) on acidic resistance of erythrocytes. Acidic erythrograms of blood after addition of: *1*) 0.001% $Pb(NO_3)_2$ solution, *2*) 0.001% $Pb(NO_3)_2$ solution and 0.1% (I), *3*) 0.001% $Pb(NO_3)_2$ solution and 0.01% (I), *4*) control

All these show that under the influence of 1-(0,0-diethylphosphonemethyl)-3,7,10-trimethylsilatrane the resistance of the erythrocytes affected by heavy metal salts to a lysing agent increases greatly due to stabilization of the erythrocyte membranes.

It has been also shown that 1-(chloromethyl)silatrane does not affect the formation of haemolysins and hemagglutinin antibodies in mice blood in response to immunization with ram erythrocytes.

4 The Effect on Endocrine Functions

The androgenic, anabolic, uterotropic, antioestrogenic, gestagenic, glucocorticoid, antiinflammatory, mineralocorticoid, antimineralocorticoid and adrenocorticoid activity of silatranes has been studied[59].

The preparation was administered per os into the animal stomach as a starch suspension at a 100 mg/kg dose.

The influence of 1-(chloromethyl)silatrane on endocrine functions regulated in the organism by steroid hormones is represented semiquantitatively in Table 12. These data show that 1-(chloromethyl)silatrane displays a considerable antigestagenic activity. On administration to preadolescent male rats for 10 days 1-(chloromethyl)silatrane activates the hypophysis-adrenal system producing increased weight of adrenal glands and decreased weight of thymus in mature rats. Besides this, the preparation induces increased weight of testes and seminal vesicles. The preparation

Table 12. The influence of 1-(chloromethyl)silatrane on the endocrine functions regulated by steroid hormones

Activity	Conclusion[a]	Activity	Conclusion[a]
Androgenic	–	Thymolytic	+
Anabolic	–	Antiinflammatory	–
Antiandrogenic	–	Glycocorticoid	++
Gestagenic	–	Mineralcorticoid	–
Antigestagenic	++	Antimineralcorticoid	–
Uterotropic	–	Adrenocorticoid	+
Antioestrogenic	–	Gonadotropic	–

[a] (+) Low activity; (++) pronounced activity; (–) no activity.

does not influence the gonad weight of adolescent rats but stimulates their adrenal gland function. This may be due to the fact that 1-(chloromethyl)silatrane activates the gonadotropic system of the hypophysis only when the development of the latter is completed (9–10 weeks of postnatal life when the animal weight amounts to 100–120 g).

1-(Chloromethyl)silatrane has a well-defined glucocorticoid activity (Table 13). Thus, on subcutaneous administration over four days it increases the glycogen content of the liver 3–4 times with unchanged glucose level in the blood.

Thus when the adrenal gland functions are blocked 1-(chloromethyl)silatrane normalizes carbohydrate metabolism stimulating accumulation of glycogen in the liver without any thymolytic effect.

5 Treatment of Wounds, Burns and Ulcers

Silicon is an essential element for epithelial and connective tissues to which it imparts strength, elasticity and impermeability[2)]. In these tissues silicon acts as an agent which holds proteins and acidic mucopolysaccharides together. This may explain the stimulating effect of a number of silicon compounds, especially silatranes, on the regeneration of connective tissue and, above all, collagen. These facts have led to investigations of the influence of silatranes on the processes of wound healing[60–62)]. For this purpose some scores of non-toxic silatranes favouring the healing of wounds in test animals have been studied.

The influence of these preparations on healing processes in wounds on subcutaneous, peroral or local administration as 5–10% ointments (Table 14) has been examined.

1-Alkoxysilatranes and 1-(chloromethyl)silatrane have proved to be most promising. Animals of the control groups were treated with Vishnevsky ointment, synthomycin emulsion and cygerol. The tests have been carried out with 204 rabbits having skin-fascial wounds of a size up to 600 mm^2.

Table 13. Glucorticoid and antiinflammatory activity of 1-(chloromethyl)silatrane on subcutaneous administration at 100 mg/kg per day to adrenalectomized rats (3 groups containing 20 animals each)

Preparation	Weight of animals [g]		Glycogen content of liver [mg]	Glucose level in blood [mg]	Weight of thymus [mg/100 g of total weight]	Weight of inflammatory granuloma [mg]
	Initial M ± m	Final M ± m	M ± m	M ± m	M ± m	M ± m
1-(Chloromethyl)-silatrane	136.6 ± 2.5	129.4 ± 2.8	1885.9 ± 129.4	63.9 ± 5.6	368.2 ± 16,3	38.8 ± 1
Hydrocortisone[a]	135.1 ± 1.9	108.1 ± 2.0	1864.2 ± 113.7	90.0 ± 2.6	145.2 ± 7.3	22.6 ± 0.5
Control	135.9 ± 1.8	123.9 ± 2.0	524.8 ± 44.9	62.5 ± 5.6	380.5 ± 21.0	38.6 ± 2.2

[a] Hydrocortisone was administered at 10 mg/kg per day.

Table 14. The influence of silatranes on healing of wounds in rabbits

Series of tests	preparations	Way of application	Number of animals	Area of wounds [mm²]	Time of healing [days]
I	1-Ethoxysilatrane	Subcutaneous	10	250	13
	1-Methylsilatrane		10	250	17
	Control		10	250	20
II	1-Ethoxysilatrane	Subcutaneous	10	300	15
	1-Ethoxysilatrane	Per os	10	300	16
	1-Ethoxysilatrane	Local, 5% ointment	10	300	13
	Control		10	300	22
III	1-Methoxysilatrane		6	250	14
	1-Tert-butoxysilatrane		8	250	15
	1-Isopropoxysilatrane	Local,	6	250	16
	1-Ethoxysilatrane	5% ointment	6	250	13
	1-Methylsilatrane		6	250	16
	1-(Chloromethyl)silatrane		6	250	12
	Syntomycin emulsion		6	250	17
	Vishnevsky ointment		6	250	18
	Control		6	250	20
IV	1-(Chloromethyl)silatrane		10	600	13
	1-Ethoxysilatrane	Local,	10	600	16
	Cygerol	5% ointment	10	600	18
	Control		10	600	22
V	1-(Chloromethyl)silatrane		10	300	14
	1-Ethoxysilatrane	Local,	10	300	14
	Cygerol	5% ointment	10	300	14
	Control		10	300	19

The first series dealing with 1-ethoxysilatrane compared with methylsilatrane consisted of three groups of animals (10 rabbits each). These compounds were subcutaneously administered as an aqueous solution at a dose of 50 mg/kg per day until complete healing of the wound[60)].

After 4–6 days the rabbits treated with 1-ethoxysilatrane displayed wounds of reduced size. The wounds were covered with a crust under which healing proceeded without suppuration. After the removal of the crust (on the 13th day) a movable soft scar not united with the underlying tissues was observed. At the same time, in the control animals the wounds displayed inflammation signs such as hyperemia, infiltration and swollen borders. By the 13th day the open wound neither closed nor cicatrized.

The regeneration processes were completed by the 17th day on administration of 1-methylsilatrane and by the 20th day without treatment (Table 14).

1-Ethoxysilatrane proved to be effective in treating wounds also via other routes of administration. Thus, in the second series of tests involving 40 rabbits with skin-muscle wounds of 300 mm^2 size 1-ethoxysilatrane was administered both subcutaneously and per os at 50 mg/kg doses and as 5% ointment in a 1 : 1 vaseline-lanolin base.

After 4–6 days of treatment the wound lips were covered with epithelium and were fixed and united with the wound fundus. 8–10 days after the crust removal the fundus of the wound was clean, pink and at the same level as the outer lips of the wound. The periphery of the wound presented a wide zone of epithelium.

On local use of 1-ethoxysilatrane the area of wounds diminished considerably. The wound was completely closed within 12 days after treatment. On subcutaneous and peroral administration, healing was completed in 15 or 16 days as compared to 22 days in the control.

In the third series, 1-methoxy-, 1-ethoxy-, 1-isopropoxy-, 1-tertbutoxy- and 1-(chloromethyl)silatranes were examined in comparison with well-known preparations widely used in therapy such as Vishnevsky ointment, synthomicyn emulsion, and cygerol (Table 14).

Skin-muscle wounds of 250 mm^2 size were treated with an ointment consisting of the appropriate silatrane in the vaseline-lanolin base. Inflammatory signs disappeared after treatment with silatranes, especially with 1-chloromethyl- and 1-ethoxy-derivatives. By the 8–10th day of treatment the wound has almost completely been epithelialized, but for a small spot in the centre covered with crust (Fig. 8).

1-(Chloromethyl)- and 1-ethoxysilatranes caused complete healing within 12–13 days after treatment. The wounds treated with other silatranes healed completely within 14–16 days (Table 14). In all cases a well-displayed zone of epithelialization was observed during the process of healing. In the centre of the zone one could see a small spot covered with crust which revealed, after the crust removal, a clean and pink wound fundus.

Regeneration of the wounds treated with synthomycin emulsion completed by the 17th day, that of the wounds treated with Vishnevsky ointment by the 18th day. The control animals displayed completely healed wounds not earlier than 20 days after treatment.

Thus, all the examined silatranes produce a more rapid wound-healing than Vishnevsky ointment and synthomycin emulsion.

In the fourth group the results of treatment of wounds with 5% 1-ethoxy- and 1-(chloromethyl)silatrane ointments and cygerol were compared.

Wounds with an area of 600 mm^2 were treated daily with the above preparations. The wounds treated with silatranes healed with rapid epithelialization of the wound lips. Soon they were covered with a crust under which the healing proceeded without suppuration (Fig. 9). The wounds treated with 1-(chloromethyl) and 1-ethoxysilatrane were completely healed within 13 and 16 days, respectively, and in the case of cygerol within 18 days. No complications have been observed.

Histological data have shown an early fibroblast reaction, formation of collagen fibres and differentiation of connective tissue on treatment of wounds with silatranes[61)].

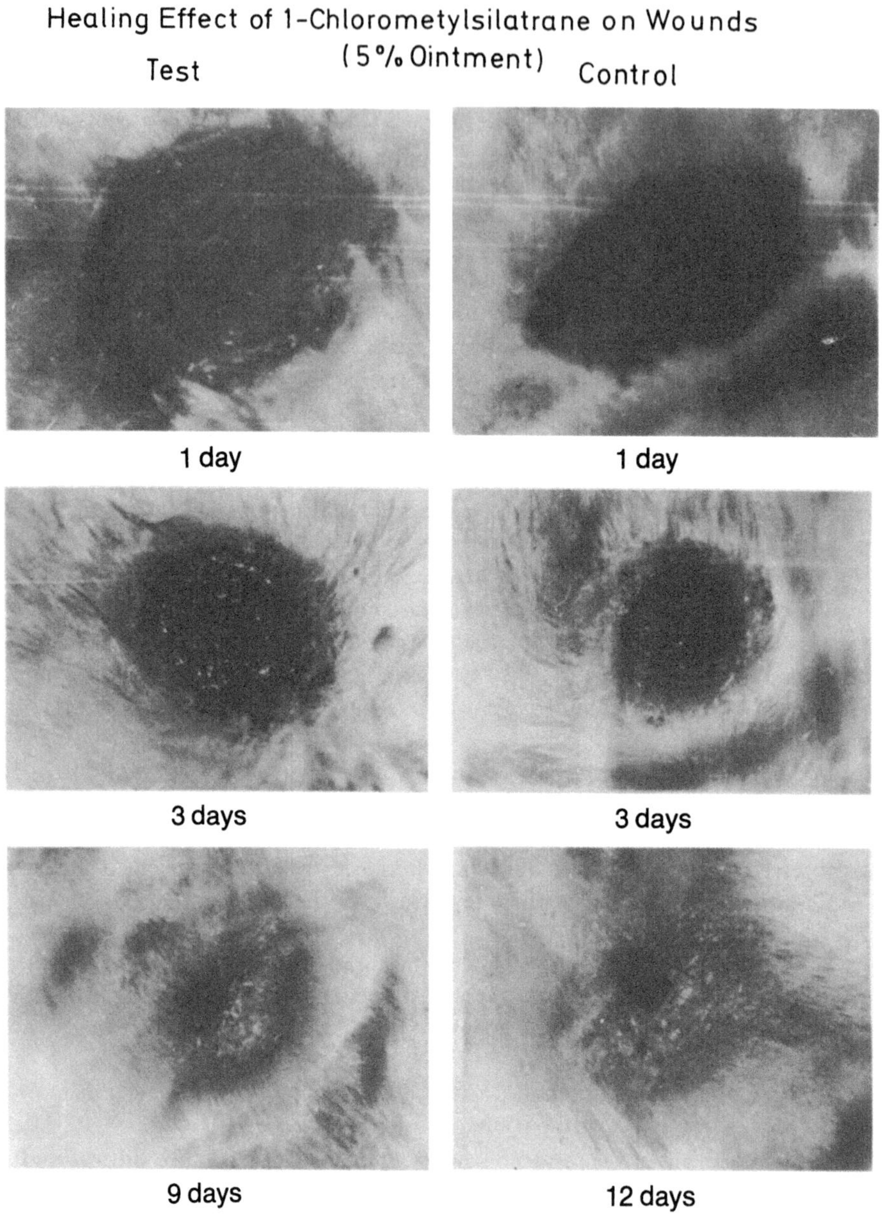

Fig. 8. Dynamics of healing of wounds treated with 5% 1-(chloromethyl)silatrane ointment (to the right – untreated wound)

An increase in the silatrane content up to 10% does not lead to a faster healing. In this case, the application of 1-ethoxy-, 1-(chloromethyl)silatrane and cygerol produces complete healing within 14 days.

Wounds treated with silatranes displayed a thicker layer of newly formed epithelium than those in the control animals and others treated with cygerol. The wounds

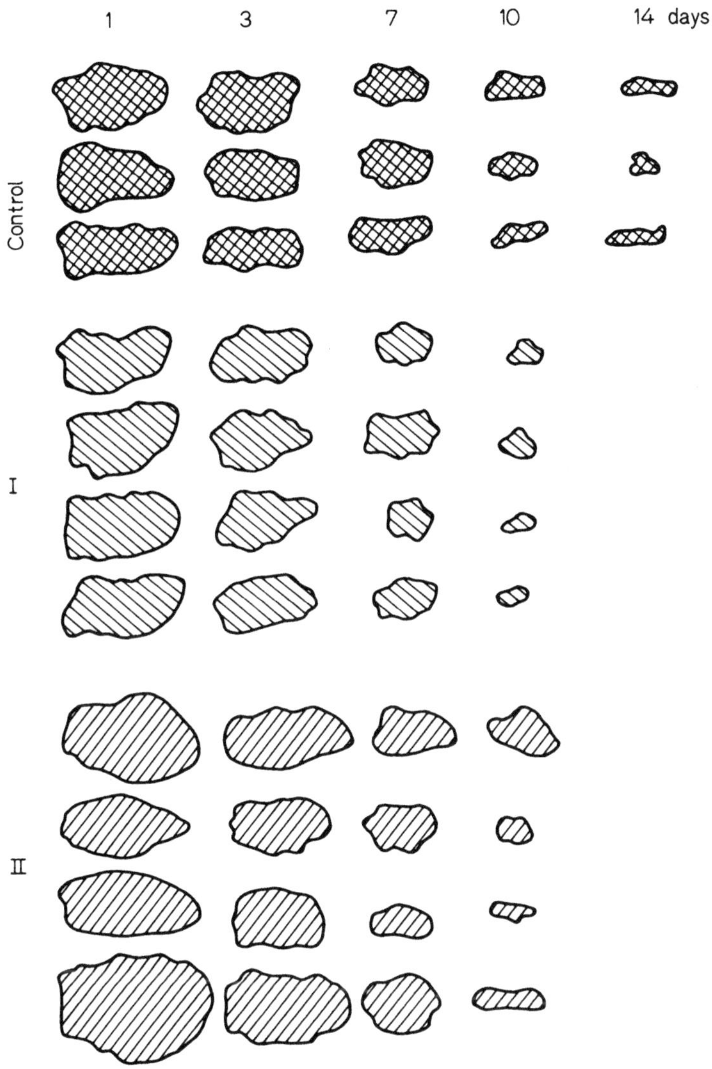

Fig. 9. Healing effect of 1-ethoxysilatrane (I) and 1-(chloromethyl)silatrane (II) on wounds

were filled in the upper layers with a newly formed connective tissue rich in blood vessels and cell elements; the lower layers contained mainly an intermediate material. The collagen fibres were arranged horizontally in fascicles of different thickness. The wounds in the control were not epithelialized completely. Differentiation of the connective tissue proceeded better in animals treated with silatranes.

The data of Table 14 show that treatment of pure wounds with silatranes is more effective than that with Vishnevsky ointment, synthomycin emulsion and cygerol.

Experimental studies of the dynamics of leucocytes and erythrocytes in healing silatrane-treated wounds were carried out using 90 male and female rabbits[65]. The hair on the back of the animal was cut off and the skin was dissected away together with the subcutaneous fatty tissue. The muscles were cut to the depth of 3–4 mm, the size of the wounds being 300 mm[2]. The wounds were treated daily with 5% 1-ethoxy- and 1-(chloromethyl)silatrane ointments and with Vishnevsky ointment, synthomycin emulsion and cygerol for comparison. The animals of the control group received no treatment.

Changes in the erythrocyte and leucocyte numbers as well as the differential blood count were examined.

During the first day after the operation the erythrocytes were reduced to 300000–500000. During the following days the erythrocyte count in the animals treated with silatranes increased, attaining the normal value within 10 days. At the same time, the erythrocyte count of the animals treated with Vishnevsky ointment and synthomycin emulsion normalized in only 15 days. The erythrocyte count in the control group animals did not reach the normal value even after 20 days.

One day after the operation the leucocyte count both in treated and untreated animals increased by 4000–5000[65]. On the 5th day the leucocyte count of the animals treated with silatrane ointments reduced to 9000 ± 500. The leucocyte count of the animals treated with Vishnevsky ointment, synthomycin emulsion and cygerol reduced to 10800 ± 100, 10200 ± 300 and 10000 ± 200, respectively. The leucocyte count of the untreated animals was 11000 ± 100 5 days after the operation. The leucocyte count of the silatrane-treated animals was normal after 15 days, i. e. by the time of complete healing of the wound, in the untreated animals not even after 20 days.

The most characteristic change in the differential blood count is an increase in the lymphocyte content by 8–10% in both treated and untreated animals. In all the treated animals the lymphocyte count reduced to a normal level by the time of complete healing. The untreated animals had normal lymphocyte count considerably later.

Thus, healing of wounds treated with silatranes is accompanied by a normal reaction of the organism which is expressed by rapidly progressive anaemia and moderate leucocytosis. However, in this case all the processes advance more quickly. Furthermore, the silatranes studied suppress inflammatory processes. This is proved by the normal leucocyte content by the time of wound healing.

The influence of silatranes such as 1-methyl-, 1-ethoxy-, 1-(chloromethyl)silatrane and 1-(chloromethyl)homosilatrane on healing of wounds has been studied using 260 white rats[63].

The hair of the back of the head was mechanically removed from an area of 3 x 4 cm under ether anaesthesia. The skin was dissected to the fascia. After the operation the wound was daily treated with a 5% ointment consisting of the appropriate silatrane and a vaseline-lanolin base until complete microscopic healing. The wounds of the control animals were covered only with the vaseline-lanolin mixture. The other control group was not treated at all. The wounds were examined daily and measured until complete healing. Then the animals were killed. From the wound area a piece of skin was taken for a histological analysis.

The wounds treated with 5% 1-methyl- and 1-(chloromethyl)silatrane ointments healed simultaneously within 28 days. The same healing time was observed for the wounds of the control group. A microscopic study of internal organs of the silatrane-treated animals has not revealed any visible pathology[63].

The influence of 1-isopropoxylsilatrane on the proliferative tissue has been studied[64]. According to the histological data, the effect of the 0.5% 1-isopropoxysilatrane ointment on the development of the granular-fibrous tissue is not limited by purely quantitative stimulation. This silatrane seems to influence the molecular arrangement of the granular-fibrous tissue that makes the morphogenetic effect on the course of epithelialization more rapid and strong.

During the investigation of the effect of intraperitoneal administration of silatrane at a dose of 150 mg/kg per day on the development of histamine-induced experimental ulcers in white rats there have been identified 10 compounds of this class which produce a strong antiulcerous action[67]. The antiulcerous effect of these compounds is manifested in a lower number of ulcerous injuries and a smaller edema of the mucous membrane of the stomach. The silatranes appear to produce both protective and preventive effects on the development and course of histamine-induced ulcerous process in the stomach.

Sulfur containing salts of general formula

$$[RR'S(CH_2)_mSi(OCHXCH_2)_n(OCH_2CH_2)_{3-n}N]^+Y^-$$

where R,R' = alkyl, aralkyl; Y = I, Br; m > 1; X = H, CH_3; n = 0–3 belong to one of the most promising types of silatranes in this respect[68]. The antiulcerous effect of these salts has been studied on white rats. The animal were given only water for 48 hours. Then they were intraperitoneally injected with aqueous solution of the appropriate thionic salt (LD_{50} = 0.1 mg/kg) and 30 minutes later with aqueous histamine solution (10 mg/kg). After 24 hours the animals were decapitated and autopsied. The stomach was examined and the number of ulcers counted.

The examination has shown that all studied compounds do produce both protective and preventive effects on the development and the course of the histamine-induced ulcerative process occurring in the mucous membrane of the stomach. The average number of ulcers in the control group was 8.1 as compared with 2.3–0.5 in the tests.

Clinical tests of the 5% 1-(chloromethyl)silatrane ointment have been carried out in four hospitals in the Soviet Union.

In one of these hospitals (Leningrad) two groups of patients having wounds of different origin were involved. The patients of the first group were treated with 1-(chloromethyl)silatrane ointment placed directly onto the wound surface or onto a sterile dressing. The dressings were changed daily or every second day till complete healing.

The patients of the second group were treated with Vishnevsky ointment, synthomycin emulsion, anaesthezine ointment and sterile vaseline.

Examination of suppurating wound microflora has shown that in both groups the suppuration process was mainly caused by staphylococci which are resistant to or had low sensitivity to penicillin and streptomycin.

Beside the clinical observations the patients have been examined for the amount of haemoglobin and erythrocytes, the colour index of the blood, the amount of leucocytes or blood count, erythrocyte sedimentation rate, total protein, residual nitrogen and bilirubin of the blood as well as the specific gravity and microscopic morphological components of the urine. The parameters were measured in order to determine the effect of 1-(chloromethyl)silatrane on the organism as a whole. The patients of both groups did not exhibit any considerable difference in these data. In several cases some deviations from normal values were observed in both groups, usually during the postoperative period or during the degenerative-inflammatory period of wound healing. In the regenerative period these indices were normal again.

When suppurative wounds were treated with the widely used Vishnevsky ointment or synthomycin emulsion they soon became free from necrotic tissues to form rich granulations. Application of the 1-(chloromethyl)silatrane ointment did not accelerate this process. The postoperatively silk-sutured wound healed at the proper time (7–10days).

Treatment of small abrasions, irritations, skin burns and erosions with the 1-(chloromethyl)silatrane ointment removed inflammation. Healing completed 1.5–2 times faster than on treatment with Vishnevsky ointment, synthomycin emulsion and other preparations. Healing of small wounds usually occurred under the crust to form a soft cicatrice. No complications were observed on using 1-(chloromethyl)silatrane.

It was concluded that the application of the 1-(chloromethyl)silatrane ointment in surgery is reasonable in treating small, purely skin-fascial wounds, erosions, skin irritations and burns.

In another hospital (Gorky) the clinical tests of 1-(chloromethyl)silatrane were carried out on 21 patients from 6 to 52 years of age. 14 of them had granulating wounds after thermal burns of 7–800 cm^2 size, 4 patients displayed superficial burn wounds (1–15% of the body surface), 3 patients had wounds which were not healing where donor skin had been removed. Most granulating wounds were characteristic of abundant suppurative discharge and poorly-defined marginal epithelialization.

The 1-(chloromethyl)silatrane ointment was applied along with the usual treatment. Uni- or bilayered gauzes soaked with the preparation were placed on to the wound surface 2–11 times.

The 1-(chloromethyl)silatrane ointment was also used in local treatment of 4 patients with fresh superficial burns (II–III degree). The treatment started on the first to the third day after the trauma. Under the influence of this preparation the wound epithelialized very well. A slight seropurulent discharge did not prevent epithelialization. Healing was completed within 10–15 days. The 1-(chloromethyl)silatrane ointment also proved to be rather effective in treating unhealed donor wounds formed due to suppuration in places where the donor skin had been excised. These wounds also showed very quick epithelialization.

The use of the preparation in treating old wounds with an abundant suppurative discharge was less effective. Local application of the 5% 1-(chloromethyl)silatrane ointment did not cause any side effects. It is concluded that the 1-(chloromethyl)silatrane ointment may be applied along with other preparations in local treatment of fresh suppurative burns.

At one of the hospitals in Moscow the 5% 1-(chloromethyl)silatrane ointment was used in treatment of 20 patients. Dressings with the ointment were applied daily or every second day. All the patients were cytologically examined for wound prints, inoculation of microflora, and sensitivity to antibiotics.

The patients of the first group which were treated with 1-(chloromethyl)silatrane showed more effective growth of granulations and improved tissue growth during the first 7–8 days. After this the healing rate increased, as confirmed cytologically.

The patients of the second group treated with 1-(chloromethyl)silatrane also exhibited a more rapid and uniform granulation during the first 10–12 days. The amount of suppurative discharge increased as well as the number of decomposing neutrophils (according to cytological data).

Six patients were treated alternatively with the ointment and antiseptics (potassium furagin, dioxidine) and this led to fast decontamination of wounds, an active growth of granulation and an intensive cell regeneration.

Six patients suffered from microbic eczema and dermatitis developing around the suppurative wounds. After application of 1-(chloromethyl)silatrane for three times perifocal phenomena completely disappeared.

It is concluded that the 1-(chloromethyl)silatrane ointment is effective in treating microbial eczema and dermatitis accompanying suppurative wounds. The application of this ointment in treatment of trophic ulcers and suppurative wounds has no advantage over widely used antiseptic preparations.

The 1-(chloromethyl)silatrane ointment has been studied also under clinical conditions with a group of sportsmen which had been operated for false joint of the tibia or subcutaneous rupture of Achilles tendon and suffered from post-operative dermal necrosis. The preparation was applied on dressings which were changed every second day. The injured tissues were not given any additional treatment. Already in 2–3 days some growth of the granulation tissue and epithelialization of the wound lips were observed. The formation of the crust occurred without suppuration and favoured the healing. The healing time of the wounds treated with the 1-(chloromethyl)silatrane ointment was 15–17 days. The healing was characteristic of a smooth connective tissue with no scars. No side effects such as skin irritation, allergy etc., were observed. The healing time in the control group was 22–26 days.

The ointment was also tested on the sportsmen participating in cycle racing on roads and cycling tracks. After falling off at high speed (up to 50 km/h) the surface of injured tissue amounts to 300 cm^2 (thigh, shoulder, etc.). Patients of the test and control groups had relatively similar injuries. The ointment was applied immediately after disinfecting the wounds with peroxide. The ointment dressings were daily placed before going to bed until complete healing.

The average healing time of the 1-(chloromethyl)silatrane treated wounds was 16 days as compared to 24 days for treatment with synthomycin emulsion.

It is concluded that the 1-(chloromethyl)silatrane ointment produces a stimulatory effect upon healing of wounds.

All the above observations show that silatranes are very promising in treatment of wounds and burns, especially in combination with appropriate antiseptic preparations[68a)].

6 Intensification of Hair Growth

Silicon compounds are very important for the growth of hair and nails in man, feathers in birds, and hair, horns, hoofs and claws in animals. Here silicon is functioning as an agent which holds the keratin macromolecules together and which along with disulphide bridges imparts mechanical and chemical strength and impermeability to fluids[2].

This all suggests that silatranes which may have the function of donors of silicon essential living organisms and stimulate the biosynthesis of nucleic acids, proteins and the genesis of connective tissue are able to intensify the growth of hair and horny tissue as well. This suggestion has been well confirmed by experiments carried out for this purpose.

The influence of silatranes on the growth of hair in test animals was tentatively studied in 4 series of tests. The preparations were administered per os and applied on the skin as ointments[69].

After making all necessary measurements in the 1st series of tests the animals were fed on a diet containing 10 mg/kg of 1-ethyl- and 1-ethoxysilatrane given 5 days a week for the period from April to May, 1972. The control group of animals was kept under the same conditions, on the same diet but free from silatranes.

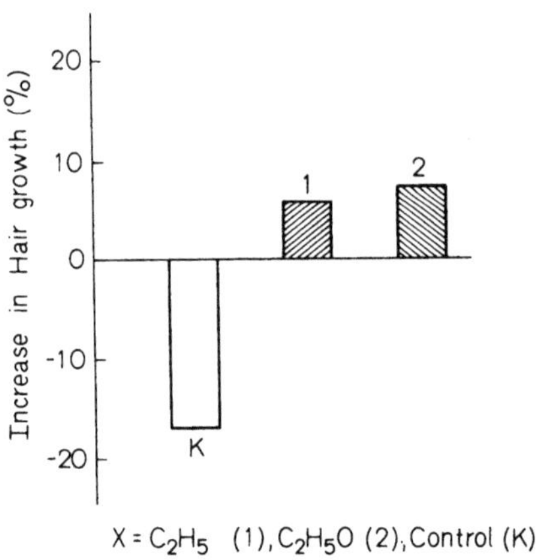

Fig. 10. Influence of silatranes, $XSi(OCH_2CH_2)_3N$ on hair growth in guinea pigs fed on the diet containing 1-ethylsilatrane (*1*) and 1-ethoxysilatrane (*2*) in spring (10 mg/kg per day; March-May, 1972) in comparison to the control (K)

Table 15. The effect of perorally administered silatranes, $XSi(OCH_2CH_2)_3N$, on hair growth in guinea pigs

X	Number of hairs [per cm^2]	Average length [mm]	Diameter of a hair [μ]	Increase in hair growth [%]
Control	376 ± 9.2	36 ± 3.2	65 ± 2.6	−17.4
C_2H_5	263 ± 9.2	30 ± 3.2	63 ± 1.6	4.7
C_2H_5O	525 ± 14.6	30 ± 2.0	72 ± 1.8	6.8

Table 16. Increase in hair growth of guinea pigs administered with silatranes, $XSi(OCH_2CH_2)_3N$, perorally

X	Control	CH_3		C_2H_5O	
Dosage [mg/kg]	0	10	50	10	50
Increase in hair growth [%]	8.9	36.0	40.5	34.4	25.0

Animals fed on the 1-ethoxysilatrane and 1-ethylsilatrane-containing diet displayed an increase in hair growth by 6.8% and 4.7%, respectively, as compared to the initial data. The animals of the control group, on the other hand, showed a decrease in the number of hairs by 17.4% (Fig. 10). No change in length and thickness of hairs was observed in any of the three groups (Table 15).

In the second series of tests (August–October, 1972) the effect of 1-methyl- and 1-ethoxysilatrane was studied. The preparations were given with food 5 times a week for two months. The increase in hair growth was 40.5% as compared to 8.9% in the control group (Table 16, Fig. 11).

In animals of the third and fourth series an alopecia was induced by application of 2% epilinic ointment. Then the hairless spot was treated with the studied silatrane as a 5–10% ointment. The best effect was observed with the 5% 1-ethoxysilatrane ointment. In this case the increase of hair growth was 55.4% (Table 17).

A higher concentration of the preparation decreases its efficacy (with the 10% ointment, the increase in hair growth was 41.4%).

Morphological investigations have shown that 1-(chloromethyl)- and 1-ethoxysilatranes produce the greatest effect of all the studied compounds.

Histological analyses were done on the skin samples taken by biopsy[69a]. Histological sections of animals fed on the 1-(chloromethyl)silatrane-containing diet showed no sharply-defined morphological changes in the epidermis.

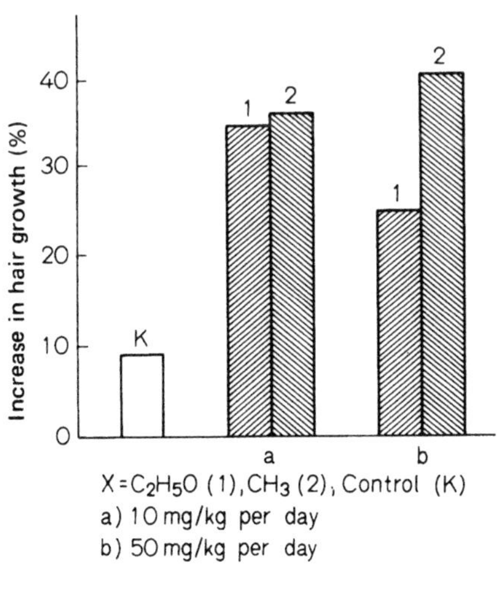

Fig. 11. Influence of silatranes, $XSi(OCH_2CH_2)_3N$, on hair growth of guinea pigs fed on the diet containing 1-ethoxysilatrane (*1*) and 1-methylsilatrane (*2*) in autumn (10 and 50 mg/kg per day; August–October, 1972) in comparison to the control (K)

Table 17. Increase in the hair growth of epilated guinea pigs on treatment with silatrane ointment, $XSi(OCH_2CH_2)_3N$

X	Control	C_2H_5		C_2H_5O		$ClCH_2$	n-$C_3H_7SCH_2$
Silatrane Content [%]	0	5	10	5	10	5	5
Increase in	10.7[a]	15.3	4.5	55.4	41.4	–	–
hair growth [%]	15.6[b]	–	–	–	–	43.3	28.6

[a] July – November 1972.
[b] May – August 1973.

Some changes were observed in the dermis. This was expressed by a changed cellular constitution of the connective tissue. Fibroblasts were predominant in the thin section of cellular elements.

There were no changes in the character of growing hair. Their main peculiarity was the appearance of two new follicles near the old one.

Other studies unfortunately dealing only with the time necessary for the cut hair to grow were carried out on female white rats and guinea pigs of the same age[70]. After having the hair mechanically removed the animals were fed on the diet containing 10 mg/kg of the examined preparation.

1-Isopropoxysilatrane proved to be most effective. On local application of this preparation the hair regained its initial length within 23 days (Fig. 12). The precursor of 1-(chloromethyl)silatrane, (chloromethyl)triethoxysilane, was less effective.

The hair of the control group animals reached its initial length within 25 days, that of the animals fed with 1-(chloromethyl)silatrane and 1-isopropoxysilatrane – within 18 days (Fig. 12). 1-Ethoxysilatrane did not influence the hair growth (as compared to the control).

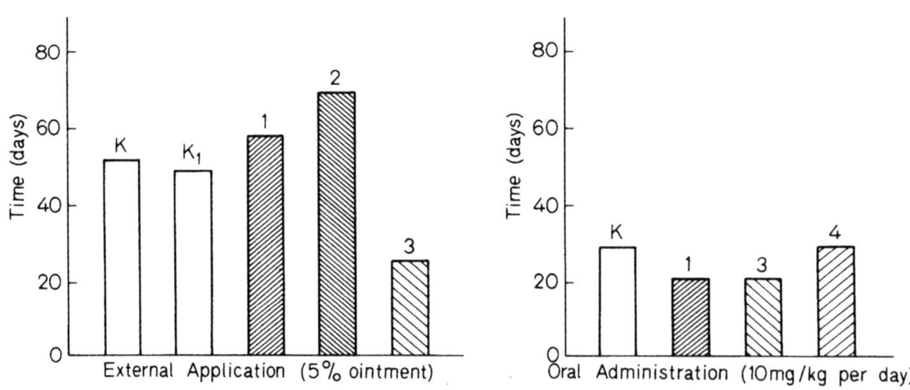

Fig. 12. Influence of silatranes on regeneration of hair in test animals. K = control, K_1 = vaseline-lanolin. *1*) 1-(chloromethyl)silatrane, *2*) 1-methylsilatrane, *3*) 1-propoxysilatrane, *4*) 1-ethoxysilatrane.

Fig. 13. Guinea pig fed on 1-(chloromethyl)silatrane-containing diet (10 mg/kg per day during 10 months)

All the data obtained have shown that some silatranes, especially 1-ethoxy-, 1-isopropoxy- and 1-(chloromethyl)silatrane effectively intensify the hair growth in animals without producing any side effects (Fig. 13). This suggested that they should be tried as possible medicinal preparations in treatment of premature hair loss (alopecia).

Extensive clinical tests of 1-(chloromethyl)silatrane and 1-ethoxysilatrane as possible drugs for treatment of alopecia have been carried out in several hospitals in Moscow, Kazan and Irkutsk[71–74]. The silatranes were used as 3–5% ointments, creams, and liniments on different, specially chosen bases.

The growth of new pigmented coarse hair was already observed after 3–4 procedures, i.e., within 10–14 days. This was manifested as a symptom of "prickly hair". The new hair was darker than before. After 3–5 courses of treating alopecia with 1-(chloromethyl)silatrane the number of anagenic hairs increased to 35%, and after 20–25 treatments to 75%. During the whole course of treatment the follicles of the hairy part of the head were actively changing to an anagenic cycle.

The treatment with 1-(chloromethyl)silatrane ointments and creams are most effective for the patients suffering from premature alopecia with or without seborrhoea. The most effect is observed in treatment of bald men. During the first course of treatment all the patients had a positive trichogram dynamics. In all cases new coarse pigmented hair appeared. The hair became more alive and dark and could be pulled out by forceps only with difficulty. No side effects were observed. Treatment of premature alopecia with 1-(chloromethyl)silatrane may be recommended for men and women, preferrably under 50 years of age. External application of this preparation may be prescibed for treatment of seborrheic baldness, alopecia areata, physiological loss of hair, moniletrex, and some other diseases causing a decrease of hair growth and hair loss. The treatment was most successful after 2 or 3 courses.

On treatment of children suffering from different types of alopecia an intensive revival of the follicular apparatus was observed in almost all cases within 5 days to 2 or 3 months.

This was followed by the appearance of fluffy or normal hair, more frequently in several places on the head, and the appearance of brows and lashes.

The best results were achieved in treatment of alopecia with liniments consisting of 3–4% of 1-(chloromethyl)silatrane or 1-ethoxysilatrane, 50% of dimethylsulphoxide and 46–47% of castor oil[74].

Dimethylsulphoxide favours good solubility of the preparations. Moreover, dimethylsulphoxide facilitates penetration of the silatrane through the skin by transporting it to the bulbs and roots of the hair. The castor oil as a component of the liniment has some advantages. It is antiseptic, contains the provitamins A and D, and can be kept for a long time. The skin was oiled, but not rubbed, with the liniment once or twice a day during 2 or 3 months. If necessary the course of treatment was carried out again 1 or 2 months later. Therapeutic efficacy of 1-(chloromethyl)- and 1-ethoxysilatrane liniments was almost the same.

All patients treated with the silatrane liniments earlier underwent unsuccessful treatments with various preparations.

At first the liniments with low concentrations of 1-(chloromethyl)silatrane and 1-ethoxysilatranes were used. Their concentrations were gradually increased to 4 and 7%, respectively.

Of 15 patients suffering from alopecia areata and treated for a long time (up to one year), 9 patients showed completely regenerated hair, 4 patients had adequately regenerated hair (the hair was long but thin). The treatment was not effective (growth of fluffy hair) in only 2 patients.

After treatment of 19 patients suffering from alopecia areata of tape form (of these 12 patients were treated for about 6 months, 3 patients for up to one year, and 3 patients for one and a half year) complete regeneration of hair was observed in 13 patients, considerable improvement in 5 patients and no visible results in 1 patient.

Of 34 patients with subtotal and total alopecia (duration of the treatment was 6 months for 16 patients, one year for 6 patients and one year and a half for 12 patients) 18 patients displayed complete regeneration of hair, 11 patients – growth of fluffy hair. No hair growth was observed for 5 patient.

After treatment of 39 patients suffering from seborrheic alopecia (duration of treatment 6 months for 22 patients, 1 year for 14 patients and one and a half year for 3 patients) complete regeneration of hair was observed in 11 patients, partial regeneration in 19, growth of fluffy hair in 8 patients. No therapeutic effect was observed for 1 patient. In all cases the treatment prevented further loss of hair.

In general, treatment of 107 patients suffering from alopecia of the above types with 1-(chloromethyl)- and 1-ethoxysilatrane liniments led to complete regeneration of hair in 34.7% of the patients, partial regeneration of hair in 43%, growth of fluffy hair in 19.6% and no hair growth in 2.8% of patients. In other words, the treatment was effective in 77.7% of patients. The use of the liniments was most successful for treatment of patients suffering from alopecia areata where a positive effect was observed in 31 patients out of 34 (91.1%).

Treatment of seborrheic alopecia with 1-(chloromethyl)- and 1-ethoxysilatrane liniments not only strengthened the roots of the remaining hair, but caused adequate regeneration of lost hair in 76.9% of the patients.

Treatment of subtotal and total alopecia with 1-(chloromethyl)silatrane liniments was successful in 64.4% of the patients.

The results of treatment of patients suffering from different types of alopecia with 1-(chloromethyl)- and 1-ethoxysilatrane liniments are shown in Fig. 14 (a, b, c, d).

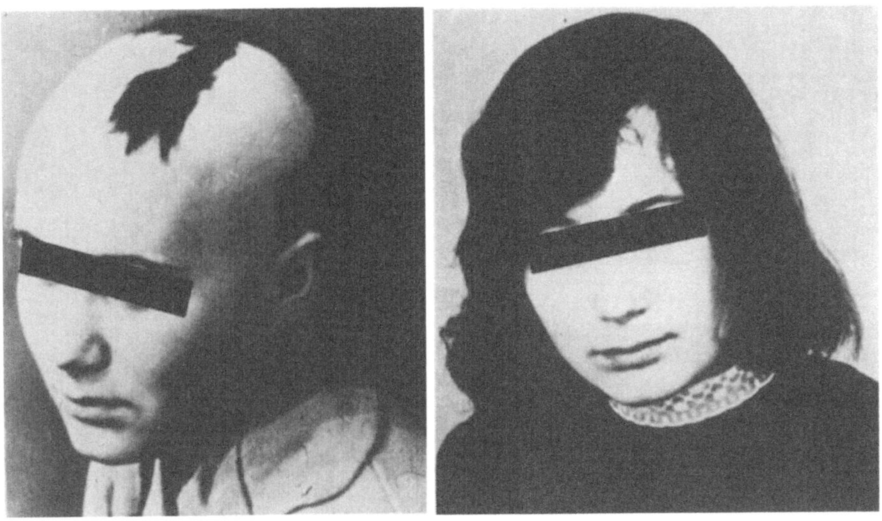

Fig. 14a

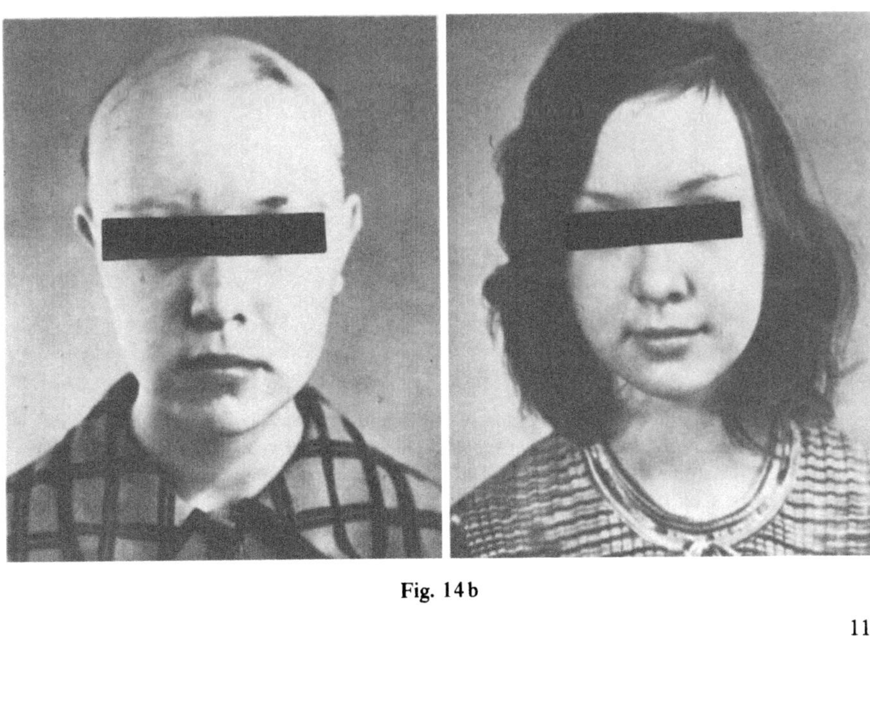

Fig. 14b

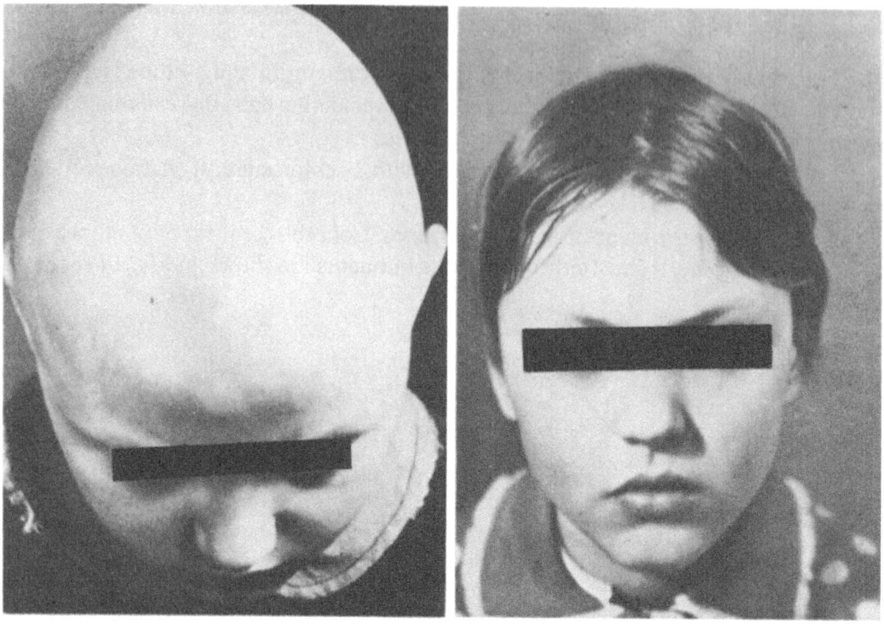

Fig. 14c

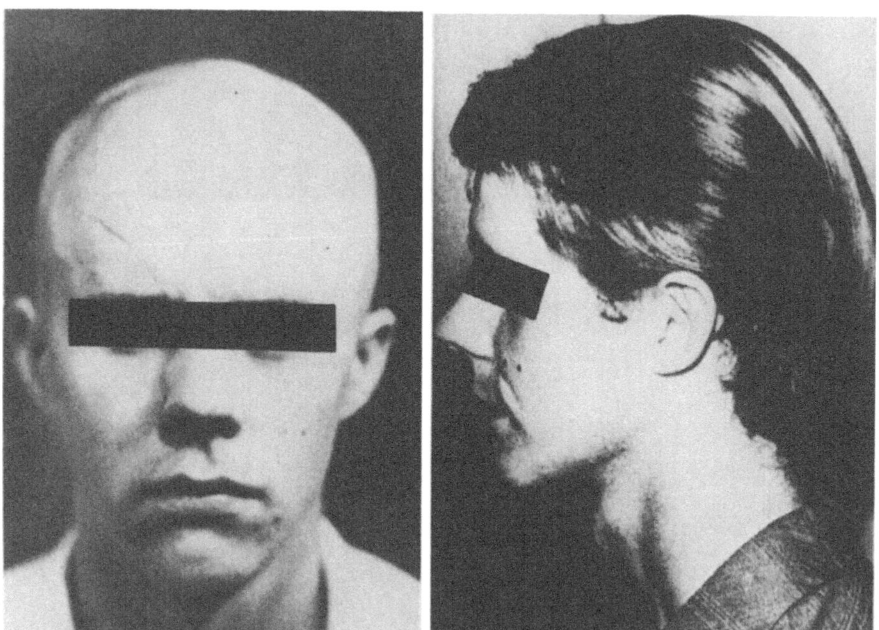

Fig. 14d

Fig. 14. Treatment of alopecia with 1-(chloromethyl)- and 1-ethoxysilatrane liniments: **a)** Patient S. P., 17 years of age, suffering from subtotal alopecia for 10 years, after treatment with 1-(chloromethyl)silatrane for 2 years; **b)** Patient I. I., 11 years of age, suffering from alopecia areata for 2 years, after treatment for 5 months; **c)** Patient D.I., 9 years of age, suffering from total alopecia for 4 years, after treatment with 1-ethoxysilatrane for 1.5 years; **d)** Patient I. V., 20 years of age, suffering from total alopecia for 3 years, after treatment for 6 months. The photograph is taken 11 months after the beginning of treatment

The patients treated with the above preparations showed no pathological changes. Some of them, on the contrary, showed better haemogram, proteinogram, Weltmann reaction and sialic acid contents.

In six cases recurrence of illness was observed but it was eliminated by repeated application of these preparations.

Thus, the use of 1-(chloromethyl)- and 1-ethoxysilatranes for treatment of patients suffering from alopecia of different types has proved to be effective and these preparations may be recommended for wide application in dermatology and cosmetology.

7 Antitumorous Activity and Immunostimulating Effect

As early as the beginning of our century it was discovered that the connective tissue may help the organism in its fight against tumour growth[75–77]. The strong protective reactions of the organism against the development and growth of tumours are conditioned by a number of factors such as the intensive development of stroma containing neutral polysaccharides, the presence of proteolytic enzymes containing lymphoid-cellular elements and plasmalytic cells.

This belief in the effective protective role of connective tissue started in the year 1970 an extensive search for new antitumorous agents among silatranes which could stimulate the formation of connective stroma[78]. It was suggested that the introduction of certain silatranes into the organism would intensify its resistance to the development of tumours. This suggestion was confirmed experimentally.

The experimental investigation was carried out on breedless white rats which were daily injected intraperitoneally with a 250 mg/kg dose of a certain silatrane within 5 days. On the first day all the animals were inoculated with Waker carcinoma. Then the silatrane was introduced ten days later except in the control group. On the 11th day the animals were killed. The tumours and the internal organs were examined pathomorphologically.

Histological, histochemical and fluorescence microscopic data have shown that after administration of a number of silatranes the tumours contain micronecroses, distrophically changed tumorous cells, and there was synthesis of collagen. 1-Ethoxysilatrane did not produce any toxic effect on the internal organs. Neither fatty degeneration of liver cells, nor spleen infiltration by lymphoid-cellular elements were observed. The tumorous tissue was characterized by extensive necrosis and proliferation of the connective tissue along the tumour periphery in which there is collagen formation. Sometimes blocks of collagen were observed in the central parts of the tumour (Fig. 15). So-called vital cells of tumorous parenchyma with pyknotic nuclei did not display the structure usual for Walker tumour; they rather resembled the cells of Ehrlich ascites tumour. Inhibition of the tumour growth exceeded 50%. The data are statistically reliable.

So it has been found out that 1-ethoxysilatrane stimulates the formation of collagen and intensifies the development of connective stroma, thus inhibiting the growth of tumorous parenchyma. The most valuable feature of this preparation is

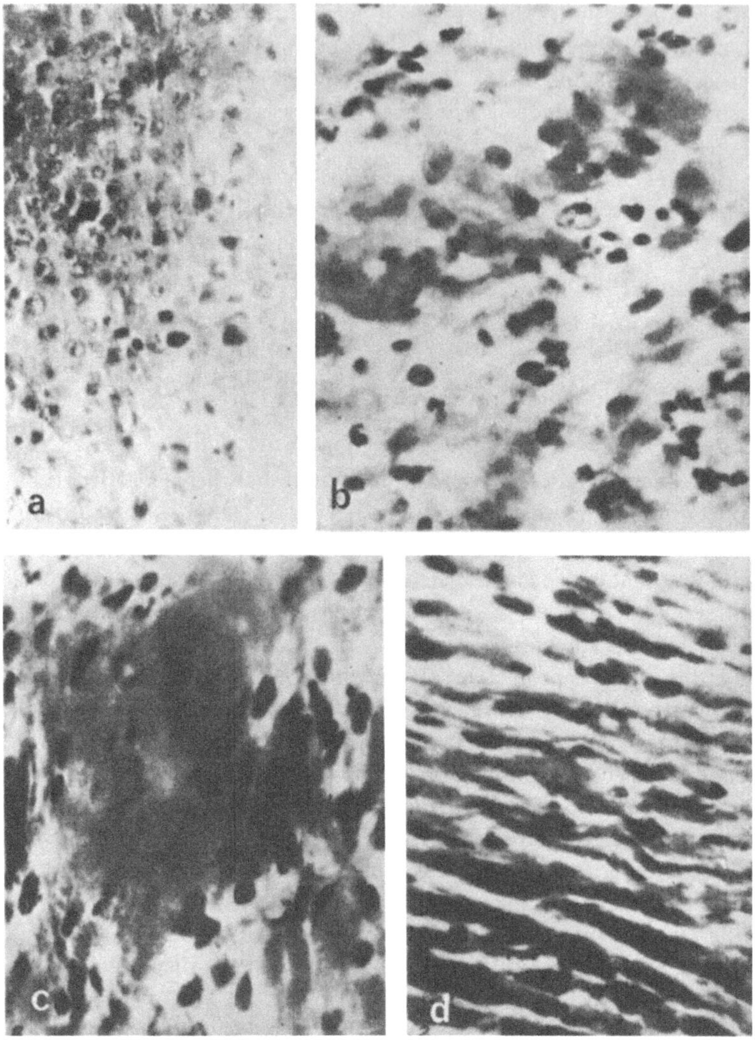

Fig. 15. The antiblastic effect of 1-ethoxysilatrane: **a)** tumorous tissue of Walker carcinoma (control); **b)** lymphocytes,fibroblasts and collagen blocks; **c)** larger collagen blocks; **d)** connective tissue capsula surrounding tumor cells

its non-toxicity, which makes it possible to obtain the desirable results without affecting healthy organs and tissues.

Of all the compounds examined (Table 18) 1-methoxysilatrane is most active. It inhibits the growth of tumours by as much as 59–79% for Walker sarcoma, 67–76% for Pliss lymphosarcoma, 47–74% for sarcoma AK and 30–55% for sarcoma 180. The average activity of 1-hydrosilatrane is 30–55% for sarcoma AK and 25–68% for sarcoma 180.

In addition, a combined effect of phthorafurum and 1-ethoxysilatrane was studied. Phthorafurum and 1-ethoxysilatrane were used at doses of 150 and 200 mg/kg

Table 18. Antitumorous activity of silatranes of the type $XSi(OCH_2CH_2)_3N$

X	Dose, [mg/kg per day]	Inhibition [%] Jensen sarcoma	Walker carcino-sarcoma	Pliss lympho-sarcoma	Carcinoma NK	Sarcoma AK	Sarcoma 180
CH_3O	210		59	67		74	52
	150		78	76		45	14
	140	18					
	100	16					
	230				53		
	460				60		
	200				64		
CH_3CH_2O	100		84	8		41	
	150		76			12	
	50		68	34		39	
	94	85	47	9	64	2	
	188	81	66	24		64	
$(CH_3)_2CHO$	313		27				
	43	89	75	9		21	39
	86	51	72	26		26	40
$(CH_3)_3CO$	143		21				
H	15					30	68
C_6H_5	0.4						40
CH_3	50		53	12			
	100		100	4			26
	150		96	12			
	200			10			39
$I^{-}\,{}^{+}[(C_2H_5)_3NCH_2$	21	66	73			66	38
	42	21	64			73	38
$I^{-}\,{}^{+}[\overline{CH_2(CH_2)_4N}(C_2H_5)CH_2]$	15					20	55
	25					4	36
	33					27	55
	40					10	62
CH_3[a]	70					15	80
	100					12	78
	150					27	0
	200					40	50

[a] 1-Methylcarbasilatrane.

Table 19. Antitumorous activity of carbofunctional derivatives of silatrane $XSi[(OCH_2CH_2)_2/(OCHRCH_2)]N$ and homosilatrane $XSi[(OCH_2CH_2CH_2)/(OCHRCH_2)_2]N$

	X	R	LD_{50} [mg/kg]	Dose [mg/kg]	Prolongation of life [%] Ehrlich Ascites tumour	Sarcoma 37	Leucaemia 5178	Lewis Carcinoma of lungs	Adenocarcinoma 755
I	C_2H_5	CH_2Cl	100	17	45	0	–	–	–
				28	35	0	–	–	–
				45	20	–20	–	–	–
				60	18	–45	–	–	–
II	CH_2=CH	CH_2Cl	30	10	0	–	–	–	–
				17	10	–	–	–	–
				28	40	–	–	–	–
				36	–45	–40	–	–	–
				0.22	10	–	0	–	–
III	C_6H_5	CH_2Cl	3.4	0.36	0	–	0	–	–
				0.60	10	–	0	–	–
				1.00	–45	–	0	–	–
IV	C_6H_5	CH=CH_2	3.0	0.22	0	–	38	–	–
				0.36	0	–	10	–	–
				0.60	18	–	–10	–	–
				1.00	–25	–	–15	–	–
V	C_2H_5O	CH_3	100	22	–10	25	0	–	–
				36	0	–10	0	–	–
				60	30	15	0	–	–
				100	–15	–10	0	–	–

VI		H	—	10	0	—	0	—	—
				17	0	—	0	—	—
				28	−40	—	0	—	—
				6	0	—	—	—	—
VII	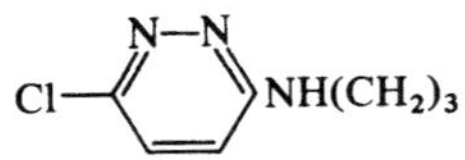	H		10	0	—	0	10	—
				17	0	—	−20	20	—
				28	0	—	0	−20	—
VIII	$CH_3CON(C_4H_9)CH_2$	H	—	45	0	—	0	—	—
				75	0	—	−55	—	—
				125	18	—	−50	—	—
				3	0	—	0	15	—
IX	$CH_3CONH(CH_2)_3$	H	—	6	0	—	0	0	—
				10	0	—	0	−20	—
				17	0	—	0	0	—
X	$ClCH_2$[a]	H	—	10	50	—	—	—	−15
				17	10	0	—	—	30
				28	0	0	—	—	0
				45	0	15	—	—	0
XI	C_6H_5[a]	CH_3		6	10	—	0	0	—
				10	−10	—	0	20	—
				17	0	—	0	25	0
				28	0	—	0	0	0

[a] Derivatives of homosilatrane.

per day, respectively. 24 hours after transplantation of Jensen sarcoma, white rats were injected 10 times with phthorafurum. After administration of phthorafurum this group of animals was injected 5 times with 1-ethoxysilatrane. This caused an almost twofold increase of the antitumorous effect of phthorafurum. A similar phenomenon was observed with Walker carcinoma (phthorafurum was injected subcutaneously at doses of 150 mg/kg and simultaneously 1-ethoxysilatrane was intraperitoneally injected at doses of 200 mg/kg per day).

These studies were followed by investigation of antitumorous activity of carbofunctional derivatives of silatrane and homosilatrane[79, 80]. The results obtained are listed in Table 19. The criterion of efficacy was life span of the test animals as compared to the untreated ones. The observations lasted 60 days.

Lymphoid leucaemia L-1210 and P-388, melanoma B_{16} and haemocytoblastosis La are not sensitive to the studied compounds.

Of compounds I-XI only 1-ethyl-3-chloromethylsilatrane (I) has a medium antitumorous effect on Ehrlich ascites tumour, the lifespan of the animals increases by as much as 35–45% (Table 19).

The replacement of an ethyl group at the silicon atom by a vinyl or phenyl group [compounds (II, III)] completely inhibits the antiblastic properties and increases the toxicity. Introduction of a vinyl group instead of a methyl group into position 3 of the silatrane skeleton [compound (IV)] retains the slight effect on the development of Ehrlich ascites carcinoma and produces an antiblastic effect on the lymphoid leucaemia L5178 y. 1-(N-Aziridinoethyl)silatrane (VI) containing an active antiblastic ethyleneimine group displays no antitumorous effect. No antiblastic effect was found in compound (VII) either.

Acetamidalkylsilatranes (VIII, IX) do not show any significant antiblastic effect. However, according to revised data[80a)], 1-(3′-acetamidopropyl)silatrane provides a medium antiblastic action on the Ehrlich ascites tumour and lymphoid leucaemia L5178 (prolongation of the life by 30–65%). 1-[3′-(n-Methoxybenzilidenamino) propyl]silatrane extends the life of mice infected with the Ehrlich ascites tumour, sarcoma 37 and adenosarcoma 755 by 30–50%[80a)].

Homosilatrane derivatives (X, XI) containing a six-membered ring instead of one of the five-memberd rings do not differ in antitumorous activity from normal silatranes.

1-(Chloromethyl)homosilatrane (X) increases the life time of mice with Ehrlich ascites carcinoma by 50%. 1-Phenyl-3,7-dimethylhomosilatrane (XI) does not affect the tumour strains studied.

1-(N-Morpholinomethyl)silatrane produces the antitumorous effect on sarcoma 37 extending the lifespan of test mice by 30–100% on daily administration at doses of 6, 10 and 17 mg/kg. At the same dosage, this preparation provides the antiblastic action on the Ehrlich ascites tumour (by 30–40%) and adenocarcinoma of large intestine (25–47%). 1-(3′-N-Morpholinopropyl)silatrane shows no antitumorous activity[80a)]. 1-(1′-Tetrahydroquinolylmethyl)silatrane also exhibits the antiblastic effect[80b)]. The antiblastic activity of silatranes is not always stable. This is not the case for antitumorous compounds of other classes.

This seems to imply that silatranes do not affect directly the cell but act via immunoreactive or hormonal systems.

Peroral administration of 1-(chloromethyl)silatrane (suspension in starch) at a 20 mg/kg dose to breedless male white rats with transplantated sarcoma 45 increases the lifespan from 26 to 34 days, i.e. by 30%. Use of 1-(chloromethyl)silatrane at higher doses does not produce such an effect.

Antitumorous and immunostimulating activity of 1-(chloromethyl)silatrane and 1-(chloromethyl)-3,7,10-trimethylsilatrane have been studied too[79]. The studies have been carried out on both healthy white rats and rats with Guerin carcinoma and sarcoma 45. In addition, the following tumorous strains have been used: sarcoma MTX, sarcoma 180, cholangioma RS-1, sarcoma-1, erythroleucosis Shvetso, Pliss lymphosarcoma.

On repeated administration of the preparation to healthy rats at a 100 mg/kg dose with a break for 72 hours no toxic signs were observed. After 5 injections the weight of the animals increased by 34.6% and after 10 injections by 69.5%. The amount of leucocytes in peripheral blood increased from 8900 to 13600 per mm^3 after 5 injections and remained at this level until the injections were stopped. (The preparation was injected per os as a suspension in 1% starch paste).

1-(Chloromethyl)silatrane did not produce any adequate effect on the complement activity of blood serum. At the same time, the preparation caused some increase in the skin test index with trypan blue (from 2.2 to 4.5 in average). This indicates an increased activity of cellular elements and fibrous structures of the skin in phagocytizing colloidal dyes.

1-(Chloromethyl)silatrane does not induce any clear autoimmunization of the organism. Thus, in the Klemparskaya reaction the value of antibodies produced by immunocompetent cells to their own erythrocytes (which is defined as the ratio of the number of patchforming cells to the total number of nuclei-containing cells) was 3.8% after 5 injections of 1-(chloromethyl)silatrane and 1.6% after 10 injections, the normal value ranging from 0 to 3%.

Repeated administration of 1-(chloromethyl)silatrane at a dose of 200 mg/kg started 8 days after the transplantation of tumour caused inhibition of Guerin carcinoma growth by 31.3%. The inhibition of sarcoma 45 growth was 14% lower. Administration of 1-(chloromethyl)silatrane to mice with Guerin carcinoma 12 days after transplantation failed to inhibit an intensive growth of neoplasms and to prevent the death of the animals. However, the average lifespan of the animals treated with 1-(chloromethyl)silatrane was 6.3 days longer than in the control group.

A daily increasing dose of the preparation did not produce any strong inhibition of the tumour growth. This fact together with the data on low antitumorous activity of the preparation and its stimulating effect on some reactions of the connective tissue system suggest that 1-(chloromethyl)silatrane does not produce a direct inhibiting effect on the growth of tumours. The observed inhibition of test neoplasms under the influence of 1-(chloromethyl)silatrane may have resulted from the stimulation of some protective reactions of the organism.

1-(Chloromethyl)-3,7,10-trimethylsilatrane (LD_{50} = 450 mg/kg, LD_{100} = 600 mg/kg) administered at 50–100 mg/kg doses to rats causes inhibition of a number of transplanted or induced tumours within the limits of 23.7–55.0%. The best-defined inhibition was observed after treatment of mice having sarcoma 180 with 1-(chloromethyl)-3,7,10-trimethylsilatrane.

The antitumorous activity of 1-(chloromethyl)-3,7,10-trimethylsilatrane does not reduce on peroral administration. The optimal doses of 1-(chloromethyl)-3,7,10-trimethylsilatrane producing the strongest antitumorous effect are 50–100 mg/kg. Unlike cytolytic preparations, an increase in the dose of 1-(chloromethyl)-3,7,10-trimethylsilatrane did not increase the inhibition of the tumour growth.

The antitumorous effect of 1-(chloromethyl)-3,7,10-trimethylsilatrane seems to be associated with stimulation of protective antitumorous reactions.

In order to study the immunostimulating effect of 1-(chloromethyl)-3,7,10-trimethylsilatrane investigations were made of the weight coefficients of immunocompetent organs (thymus and spleen), the total leucocytes and erythrocytes in peripheral blood, titre of blood complement, the amount of antigen-reactive rosula-forming lymphocytes in spleen and thymus, microbial dissemination of skin as well as skin test with trypan blue.

After the complete course of treatment with 1-(chloromethyl)-3,7,10-trimethylsilatrane at doses of 50 and 100 mg/kg per day white mice with sarcoma MTX and rats with implantated tumours displayed increased weight coefficients for spleen and thymus and a greater amount of functionally active rosula-forming lymphocytes.

The immunostimulating effect of 1-(chloromethyl)-3,7,10-trimethylsilatrane administered at 50–100 mg/kg doses to healthy animals is manifested not only by an increase in the weight coefficients and the RFL level of thymus and spleen but also by an enlarged total amount of leucocytes in peripheral blood (on an average from 9700 to 14300 per mm^3) and a slight increase in the positive function of cellular and fibrous structures of skin.

A decrease in a daily dose to 20 mg/kg is followed by a reduced immunostimulating effect of 1-(chloromethyl)-3,7,10-trimethylsilatrane. An increase in the dose to 250 mg/kg leads to a considerable lowering of the weight coefficients of thymus as compared to the control.

The above data permit the conclusion that 1-(chloromethyl)silatrane and 1-(chloromethyl)-3,7,10-trimethylsilatrane show a certain immunostimulating effect. This effect is displayed by an increase in leucopoiesis and a number of other immunological factors and seems to favour the inhibition of the growth of some experimental neoplasms.

The study of the antitumorous effect of low-toxic silatranes not affecting healthy organs and tissues and producing a certain antiblastic effect in the tests with animals indicates the necessity of further investigation of this class of organosilicon compounds in order to develop a new type of anti-cancer preparations.

8 Neurotropic Effect

A search for new types of psychotropic preparations among silatranes is rather promising considering that silatranes are donors of silicon which are digested by the organism and that some of them produce a specific effect on the nervous system of mammals.

Even the simplest 1-organylsilatranes potentiate the narcotic activity of hexabarbital[33]. They possess an analgetic and sedative effect. 1-Methylsilatrane is most active in this respect.

1-Phenylsilatrane solutions acting as local anaesthetic produce a blocking effect on conduction in the sciatic nerve of rats. At the same time, this compound even at 100 times higher concentrations does not affect conduction in the sciatic nerve of frogs.

Administration of 1-vinylsilatrane at a 100 mg/kg dose reduces the body temperature of animals by more than 2 °C within 2.5 hours[33].

1-(Chloromethyl)silatrane at higher doses (more than 100 mg/kg) produces a depressive action on the central nervous system, and displays a weak or moderate sedative effect. This effect is also manifested in a lower response to stimuli. 1-(Chloromethyl)silatrane reduces the spontaneous motor activity in mice by more than 40 times, increases the duration of hexenal sleep threefold and prolongs the arecoline tremor 1.5 times.

1-(2′-Thienyl)silatrane also markedly prolongs phenobarbital-induced sleep[44a, 44b]

The effect of a series of halosubstituted 1-alkylsilatranes on the central nervous system was studied on white mice and rats[41, 81–84].

1-(3′-Fluoropropyl)silatrane causes a strong depressive effect when applied with electrostimulus (Table 20). 1-(3′,3′,3′-trifluoropropyl)silatrane practically does not produce such an effect. (3-Chloropropyl)-triethoxysilane also induces a depressive effect on the central nervous system. However, this effect is less pronounced than with 1-(3′-fluoropropyl)silatrane.

Administration of 1-(2′-perfluorohexyl-1′-iodoethyl)silatrane to animals which have received a lethal dose of corasole increases the viability as much as 2–7 times.

1-(2′-Perfluoropropyl-1′-iodoethyl)silatrane, on the contrary, causes an opposite effect.

The study of the influence of these compounds on muscular fatigue (running track), pain stimulation (hot plate), as well as potentiation of sodium thiopental anaesthesia has shown that they both sharply lower the motor activity of the animals (Table 21).

On thermal stimulation 1-(2′-perfluoropropyl-1′-iodoethyl)silatrane produces the strongest effect. At the same time, 1-(2′-perfluorohexyl-1′-iodoethyl)silatrane increases the anaesthetic effect of sodium thiopental almost 5-fold.

Table 20. The effect of some 3′-fluorosubstituted 1-propylsilatranes and 3-chloropropyltriethoxysilane on the induced aggression of rats

Compound	LD_{50} [mg/kg]	Dose [mg/kg]	Beginning of aggression [sec]	
			Test	Control[a]
1-(3′-Fluoropropyl)silatrane	86.0	40	258 ± 90	96 ± 32
1-(3′,3′,3′-Trifluoropropyl)silatrane	10.3	1	56 ± 19	51 ± 13
(3-Chloropropyl)triethoxysilane	1500.0	40	162 ± 38	51 ± 13

[a] Average statistical values.

Table 21. The effect of some fluorosubstituted 1-organylsilatranes and of 3-chloropropyltriethoxysilane on the motor activity of rats

	Dose [mg/kg]	Running track [min]		Hot plate [sec]		Potentiation of sodium thiopental narcotic effect [min]	
		Test	Control[a]	Test	Control	Test	Control
$C_6F_{13}CH_2CHISi(OCH_2CH_2)_3N$	4040	5.3 ± 0.6					
	2020	3.6 ± 0.7	19.2 ± 2.8	18.6 ± 1.3	9.7 ± 0.9	109.8 ± 23.1	22.5 ± 2.2
	40	1.4 ± 0.1					
$C_3F_7CH_2CHISi(OCH_2CH_2)_3N$	20	2.2 ± 0.3	24.5 ± 4.4	27.2 ± 6.6	9.4 ± 0.5	75.3 ± 9.4	22.5 ± 2.2
	10	3.1 ± 0.5					
$ClCH_2CH_2CH_2Si(OC_2H_5)_3$	190	3.5 ± 0.9	19.2 ± 2.8	–		–	

[a] Average values.

Table 22. Haematological investigation of chickens fed on 1-(chloromethyl)silatrane-containing diet

Index	Before treatment		Duration of treatment [months] 1		2		3		4		5	
	Test	Control	Test	Control	Test	Control	Test	Control	Test	Control	Test	Control
Haemoglobin [%]	8.2	8.2	8.6	8.4	9.4	8.0	10.1	9.6	11.0	9.0	10.6	9.3
ESR [mm/hour]	8.0	4.0	4.0	6.0	2.0	6.0	5.0	5.4	3.0	4.0	4.0	6.0
Erythrocytes [10^6]	2.88	2.88	2.98	2.92	3.20	2.90	3.80	3.01	4.00	2.98	4.20	3.10
Leucocytes [10^3]	18.2	18.0	17.0	17.6	28.6	21.2	32.0	22.4	26.6	18.4	30.0	20.3
Acid capacity [%]	0.330	0.326	0.384	0.342	0.388	0.342	0.396	0.368	0.400	0.345	0.410	0.362

To study the tranquilizing activity of 1-(3′,3′,3′-trifluoropropyl)-3,7,10-trimethylsilatrane its ability to potentiate the action of soporific agents, the influence on the spontaneous bioelectric activity of subcrustal formations of the brain, the protective effect against stress and the change in adrenalin and noradrenalin excretion in urine has been examined[83)].

Intraperitoneal injections of 15% ethanol to mice caused deep sleep that enabled it to be used as an anaesthetic agent.

Thus, the silatrane studied increases the time of narcotic sleep and its efficacy rises with the dosage.

1-0-Silatranyl derivatives of D-mannose, D-fructose, and D-arabinose exhibit a weak depressive effect on the central nervous system[85)]. They produce a moderate cataleptic action, do not influence the duration of sodium thiopental anaesthesia and do not remove the effect of phenaminic stereotypy in rats.

The biological effect of all these compounds does not depend on the structure of the monosaccharide linked with the silatranyl residue[85)].

This shows that the neurotropic effect of these compounds seems to be caused by the same primary product of their hydrolysis, i.e., silatranol, $HOSi(OCH_2CH_2)_3N$.

Some new compounds having an effective and specific activity on the central nervous system have thus been found among silatranes. This suggests that new types of psychotropic agents may be found among the substances of this class.

9 The Effect on Domestic Fowl

The effect of 1-(chloromethyl)silatrane and 1-ethoxysilatrane on the organism and reproduction of hens has been studied[86, 87)].

The preparation was mixed with dry food (10 mg per 1 kg of weight of live hens) and fed once a day in the morning for 30 days.

The above study has shown that 1-(chloromethyl)silatrane does not produce any toxic or harmful effect on the organism of fowl. On the contrary, the weight-gain of fowl fed on the silicon-containing diet was 2.2–2.3% higher than that of the control group. The weight of live hens was also 3.1–13.8% higher compared to the control group. Histological examination of liver, kidneys, muscles, reproductive organs, and bone tissue of the fowl fed on the silatrane-containing diet has not shown any inflammatory or degenerative changes.

Examination of the morphological composition of blood has shown that 1-(chloromethyl)silatrane produces a favourable effect on the haemopoietic function. The amount of erythrocytes, leucocytes and haemoglobin has been higher in the fowl fed on silatrane-containing diet (Table 22). Here a higher level of blood alkali reserve indicating a more intensive metabolism in the organism has been also displayed. This is also confirmed by the indices of protein, lipid, carbohydrate and phosphorus-calcium metabolism (Table 23) as well as by a high productivity and better quality of meat (Table 24) and eggs (Table 25). In the first series of tests the productivity in the test group increased by 11.3% during 6 months, in the second series by 15.9% during 10 months and in the third series by 31.2% during 7 months.

Table 23. Metabolic indices for chickens[a] fed on 1-(chloromethyl)-silatrane-containing diet

Index	Control	Test
Total protein [%]	4.0	4.3
Albumins [%]	38.0	37.0
Globulins:		
α-globulins [%]	30.3	29.2
β-globulins [%]	10.9	11.1
γ-globulins [%]	19.8	24.6
Transaminases:		
Glutamine-asparaginic [units]	33.6	34.1
Glutamine-alaninic [units]	14.7	17.7
Residual nitrogen [ppm]	370.9	470.8
Total calcium [ppm]	178.9	260.0
Phosphorus inorganic [ppm]	40.7	51.6
Calcium-phosphorus indices:		
Ca x P	72.8	134.2
Ca : P	4.4	5.0
Alkaline phosphatase [Bodanovsky's units]	6.0	16.2
Glycogen [ppm]	198.0	246.0
Sugar [%]	0.17	0.22
Total lipids [%]	1.94	22.01
Cholesterol [ppm]	800.1	827.8

[a] Chickens, 10 months of age, treated with 1-(chloromethyl)silatrane during the prelaying period.

Table 24. Chemical composition of meat of the chickens fed on 1-(chloromethyl)silatrane-containing diet

Index	Before treatment		After treatment for 5 months	
	Test	Control	Test	Control
Moisture [%]	72.0	72.1	69.1	70.1
Protein [%]	21.0	21.1	24.2	22.7
Ash [%]	1.1	1.1	1.4	1.2
Fat [%]	4.2	4.2	8.2	7.6
Calcium [ppm]	210.0	213.0	2242.6	2221.8
Phosphorus [ppm]	2208.0	2187.0	6128.0	5760.0

Table 25. Chemical composition of eggs of the hens fed on 1-(chloromethyl)silatrane-containing diet

Index [%]	Before treatment		After treatment for 5 months	
	Test	Control	Test	Control
Whole egg				
Moisture	69.9	70.1	66.4	67.2
Protein	12.2	12.2	16.7	16.3
Ash	0.9	0.9	0.9	0.9
Fat	10.6	10.5	11.9	11.1
Calcium	0.12	0.12	0.15	0.13
Phosphorus	0.52	0.52	0.55	0.53
White of egg				
Moisture	86.2	86.2	85.0	85.1
Protein	13.0	12.9	16.2	14.8
Ash	0.5	0.5	0.6	0.06
Fat	0.04	0.04	0.08	0.05
Calcium	0.03	0.03	0.05	0.03
Phosphorus	0.02	0.02	0.04	0.03
Yolk				
Moisture	50.9	50.9	48.5	49.1
Protein	14.2	14.8	16.8	15.1
Ash	0.15	0.16	1.6	1.6
Fat	22.4	22.3	32.2	23.8
Calcium	0.16	0.16	0.19	0.16
Phosphorus	0.43	0.42	0.49	0.47

The data obtained have shown that 1-(chloromethyl)silatrane produces a favourable effect on the physiological processes in the organism of fowl. It increases the viability, stimulates the function of haemopoiesis thus raising the number of erythrocytes and the amount of haemoglobin. Administration of 1-(chloromethyl)silatrane stimulates metabolic reactions in the organism thus increasing the metabolism of protein, lipids, carbohydrates and phosphorus-calcium.

Of special interest is the ability of 1-(chloromethyl)silatrane to increase the productivity of fowl. The hens fed on the silatrane-containing diet at the beginning of the period of egg-laying showed a productivity increased by 11.3%, those fed on the same diet in the period of intensive laying and before laying increased the productivity by 15.9% and 31.2%, respectively. Similar results (an increase in egg-laying by 12–18%) were obtained with 1-ethoxysilatrane.

1-Ethoxysilatrane improves also the chemical composition of eggs and meat by raising the content of both protein and mineral salts.

It may thus be concluded that 1-(chloromethyl)- and 1-ethoxysilatrane are worth using in poultry farming in order to increase the productivity of hens and the quality of eggs and meat.

10 The Effect on Insects and Parasites of Animals and Plants

The ability of silatranes to stimulate protein biosynthesis[13] has given rise to investigations of their influence on the growth, development and in particular productivity of silkworm, as the silk thread extruded by the insect is of protein nature[88–90, 93]. In these investigations the silkworms which belonged to the third group of age were fed with silatranes until the cocoon formation. For this purpose the leaves of mulberry-trees were sprayed three times a day with a 0.2% aqueous solution of 1-(chloromethyl)silatrane, 1-ethoxysilatrane and 1-(chloromethyl)-3,7,10-trimethylsilatrane. All the preparations increased the viability and silk-production of the silkworm by 16 and 21% respectively (Fig. 16).

To elucidate the mechanism of increasing the sil-productivity due to an enhancement of weight of the silk shell in the total cocoon mass, the influence of 1-ethoxysilatrane and 1-ethoxy-3,7,10-trimethylsilatrane on some physiological and morphological indices of these insects has been studied[89] (Fig. 17).

It has been shown that both compounds give rise to an increasing amount of protein and haemocytes in haemolymph, enlarging the reserve of structural and energy substances in the organism. The enlarged size of silk-producing glands indicates their increased silk-producing ability. The increase of amount of formative elements of haemolymph favours the transportation and digestion of nutrients. All this gives rise to high productivity in silkworms treated with silatranes[89].

The effect of silatranes on another type of insects, the bees has been also studied. It has been found that 1-ethoxysilatrane and 1-ethoxy-3,7,10-trimethylsilatrane increase the lifespan and the weight of body and breeches of bees and stimulate the reproductive activity of the queen thus enabling the colony mass to be increased by 17–30%[91]. The results suggest that silatranes should be applied in bee-keeping

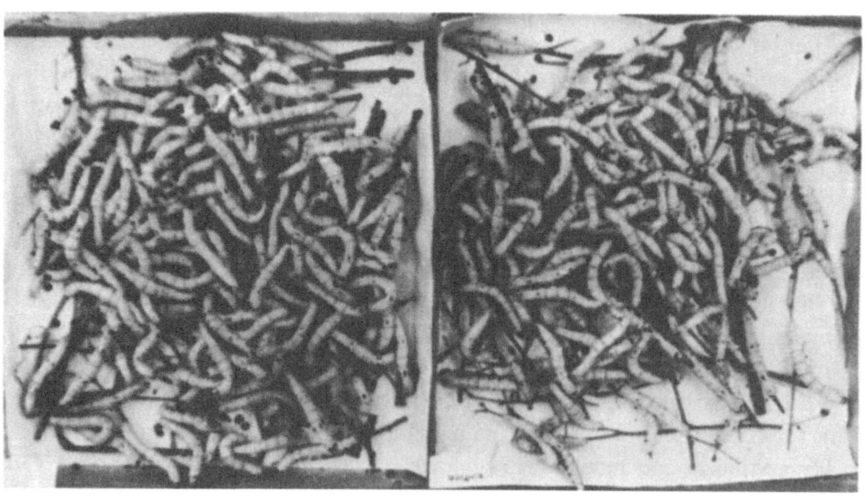

Fig. 16. Influence of 1-ethoxysilatrane on the weight growth of silkworms

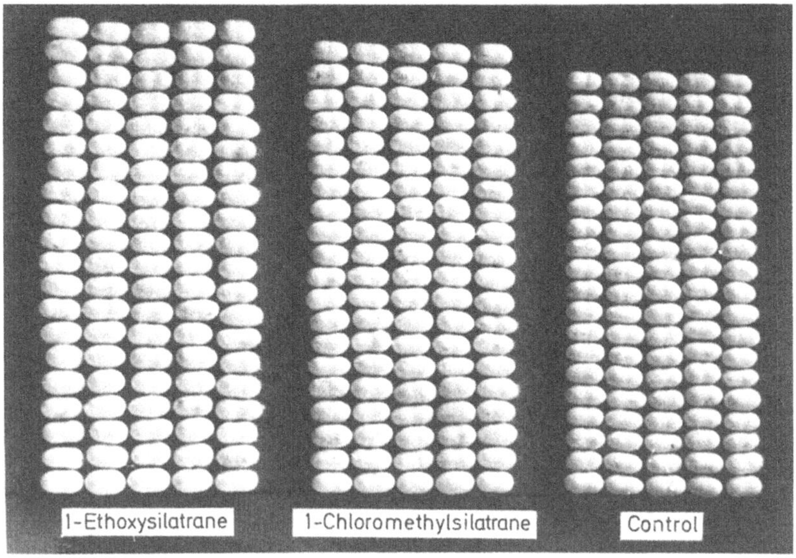

Fig. 17. Influence of 1-(chloromethyl)silatrane (*I*) and 1-ethoxysilatrane (*II*) on silk-production of silkworm

In order to investigate the mechanism of action of silatranes on insects, the effect of 1-(chloromethyl)silatrane and 1-ethoxysilatrane on the giant neuronal formations in the central nervous system of American cockroach, "Periplaneta americana", has been studied[92]. It has been found that the giant neurones of cockroaches are very sensitive to 1-ethoxysilatrane (its physiological effect is 10000 times stronger then that of 1-(chloromethyl)silatrane). The effect of both compounds is similar to that of carbomatic and organophosphorous anticholinesterase compounds.

The study of the dynamics of the action of 1-ethoxysilatrane and 1-(chloromethyl)silatrane shows that their threshold concentrations (10^{-7} and 6×10^{-4} M, respectively) induce prolonged stimulation of the nervous apparatus[92].

High concentrations (from 10^{-5} to 10^{-1} M) either quickly destroy the mechanism of biopotential generation or block the transmission in the synaptic formations, thus sharply decreasing the electrical activity of giant axons. However, the concentrations ranging from 10^{-4} to 10^{-3} M are far from physiological and this makes the interpretation of this effect rather difficult.

Numerous silatranes display an effective coccidiostatic, phyto- and zoohelminthous action.

11 The Effect on Plants

In order to study how a silatrane group which is present in phytohormones influences their activity we have investigated the action of silatranylmethyl esters of 2-methylphenoxyacetic, 4-chlorophenoxyacetic and 3-indolylacetic acids on the suspended cul-

Table 26. The effect of 2-methylphenoxyacetic acid and its silatranylmethyl ester on the growth of cultures of plant tissues

Culture of tissue		Weight of raw culture of tissue [mg/s.v.[a]] at phytohormone concentration [mol/l] a): $2\text{-}CH_3C_6H_4OCH_2COOH$ b): $2\text{-}CH_3C_6H_4OCH_2COOCH_2Si(OCH_2CH_2)_3N$					
		0	Standard[b]	10^{-6}	10^{-5}	10^{-4}	10^{-3}
Tobacco	a)	–	–	–	–	–	–
	b)	293	–	236	243	213	–
Soya	a)	25	152	35	37	143	–
	b)	44	–	46	52	224	41
Potatoes	a)	–	–	–	–	–	–
	b)	47	137	–	49	67	–

[a] Standard volume = 7.5 ml.
[b] Standard = 1-naphthylacetic acid (10^{-5} mol/l).

tures of tissues of tobacco, potatoes and soya[94–96]. The corresponding aroxyacetic acids and their esters of the type $(2\text{-}Y\text{-}4\text{-}XC_6H_3)OCH_2COOR$ where Y = H, CH_3, Cl; X = H, Cl, F; R = C_2H_5 were used as standards.

The auxin activities of the above phytohormones as well as their silatranylmethyl esters are listed in Tables 26–28.

Silatranylmethyl ester has a higher auxin activity with respect to the culture of soya tissue than free 2-methylphenoxyacetic acid. At the same time, it has a low activity with respect to tobacco and potato cultures. 1-Silatranylester of 4-chlorophenoxyacetic

Table 27. The stimulation of the growth of cultures of plant tissues under the action of 4-chlorophenoxyacetic acid and its silatranylmethyl ester

Culture of tissue		Weight of raw culture of tissue [mg/s.v.[a]] at phytohormone concentration [mol/l] a): $4\text{-}ClC_6H_4OCH_2COOH$ b): $4\text{-}ClC_6H_4OCH_2COOCH_2Si(OCH_2CH_2)_3N$						
		0	Standard[b]	10^{-8}	10^{-7}	10^{-6}	10^{-5}	10^{-4}
Tobacco (1:40)	a)	341	315	286	294	320	448	–
	b)	341	315	277	326	469	340	–
(1:100)	a)	155	345	116	130	248	423	–
	b)	155	345	161	286	426	211	–
Soya	a)	18	280	–	120	315	263	35
	b)	18	280	–	157	297	164	55
Potatoes	a)	84	246	–	94	132	376	49
	b)	84	246	–	95	123	465	107

[a] Standard volume = 7.5 ml.
[b] Standard = 1-naphthylacetic acid (10^{-5} mol/l).

Table 28. The influence of 3-indolylacetic acid (I) and its silatranylmethyl ester (II) on the growth of cultures of plant tissues and wheat coleoptiles

Concentration [mol/l]	Weight of raw culture of tissue [mg per standard volume]				Increase in growth of coleoptiles [mm]	
	Soya		Tobacco			
	I	II	I	II	I	II
0	96	96	86	86	5.3	5.3
$1 \cdot 10^{-7}$	–	–	–	–	5.9	5.7
$1 \cdot 10^{-6}$	86	341	113	156	7.6	7.4
$3 \cdot 10^{-6}$	130	327	118	234	–	–
$1 \cdot 10^{-5}$	104	177	106	212	8.5	7.7
$3 \cdot 10^{-5}$	192	113	32	190	–	–

acid and 3-indolylacetic acid (heteroauxin) display an even higher auxin activity (Tables 27 and 28).

Heteroauxin silatranylmethyl ester stimulates the growth of the cultures of tobacco and soya tissues more effectively than 3-indolylacetic acid and its ethylester. At the same time, on wheat coleoptiles heteroauxin silatranylester and 3-indolylacetic ester display almost similar effects.

In plants 3-indolylacetic acid is inactivated by 3-indolylacetic oxidase. A comparative study of the resistance of 3-indolylacetic acid and heteroauxin silatranylmethyl ester to crystalline peroxidase of horseradish (20 mkg/ml) has shown that the first compound decomposes within 10 minutes whereas the second compound does not decompose during the experiment (90 min) (Fig. 18).

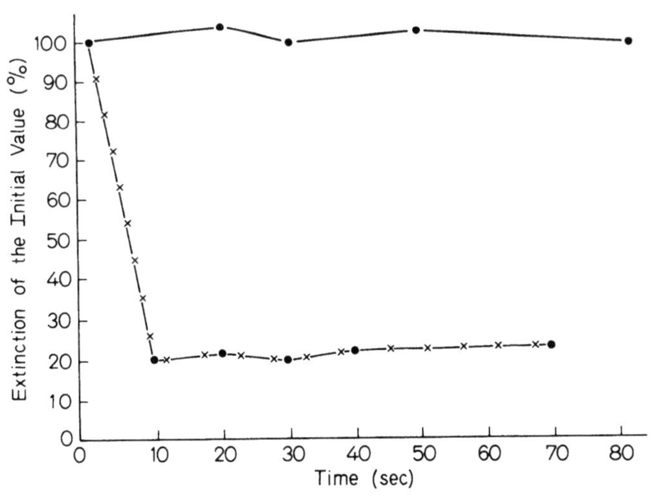

Fig. 18. Dynamics of oxidative decomposition of 3-indolylacetic acid (I) and its silatranemethyl ester (2) by horse-radish crystalline peroxidase

Table 29. The effect of 2-methyl-4-chlorophenoxyacetic acid (I) and its silatranylmethyl ester (II) on the growth of cultures of plant tissues

Concentration [mol/l]	Weight of raw culture of tissue [mg per standard volume]					
	Tobacco		Soya		Potatoes	
	I	II	I	II	I	II
0	26	26	18	18	84	84
10^{-7}	51	61	97	87	119	–
10^{-6}	192	230	140	141	262	265
10^{-5}	94	121	107	73	735	841

These data show that the higher growth-stimulating activity of heteroauxin silatranylmethyl ester in the cultures of plant tissues is due to the fact that the 3-indolylacetic acid fragment is protected from rapid decomposition by the oxidase and is able to influence growth for a longer time than free 3-indolylacetic acid[96)].

This may be confirmed by the fact that silatrane derivatives of synthetic auxins resistant to IAA-oxidase are practically as active as auxins themselves (Table 29).

Thus, the presence of a silatrane group in the phytohormone molecule prolongs its effect or facilitates the auxin transport through the biomembranes.

By extensive studies of the influence of a large number of silatrane derivatives of different structure on cultures of plant tissues several very active compounds were recognized and examined in their effects on seeds, seedlings of plants and various cultivary such as tomatoes, cucumbers, grapes, flax, etc.

The influence of 2-methylphenylacetic silatranylmethyl ester, silatranylcholine iodide and 1-(chloromethyl)silatrane on the growth and development of tomatoes has been investigated in several greenhouses in East Siberia. Treatment of tomatoes at flowering time with aqueous solutions of the above compounds has considerably increased the amount of fruits and shortened ripening time. In spite of some decrease in size the total amount and weight of tomatoes from one bush increases 1.5–3 fold.

The influence of silatranes on the growth and development of flax was investigated using aqueous solutions of 1-(chloromethyl)-3,7,10-trimethylsilatrane[97)]. The treatment of flax was carried out 10–15 days after the beginning of growth in the period of intensive formation of fibre in stalks. The best results were obtained after spraying of flax with 1% 1-ethoxysilatrane solution. The strength of hackled fibre was 13 kg/mm^2 as compared to 11 kg/mm^2 in the control. The number of long fibre indicating its quality was 13.1 for treated flax as compared to 11.5 for the control. Even in regions of drought an increase in yield of both total fibre (by 0.2–2.2%) and long fibre (by 0.4–2.8%) was observed.

The possibilities considered of application of some silatrane derivatives as specific stimulators of the growth of plants suggest that these compounds influence, in some direct or indirect way, the components of the nucleic structures (such as nucleic acids, histones) constituting the genetic apparatus of the cell.

12 Other Kinds of Biological Action

On investigation of the influence of 1-(chloromethyl)silatrane on the permeability of cutaneous tissues of rats it has been found[98] that this compound increases the permeability and, consequently, the trophicity of skin. Intracutaneous administration of 1-(chloromethyl)silatrane has proved to be more effective at a dose of 50 mg/kg than 250 mg/kg.

The effect of intratracheal administration of 1-ethoxysilatrane and 1-(chloromethyl)silatrane on experimental silicosis has been studied[99]. Far from inhibiting silicosis, the former compound even stimulates it. 1-(Chloromethyl)silatrane has proved to be a more fibrogenous compound than crystabolite, which is considered to be one of the most fibrogenic varieties of silica.

Intratracheal administration of 1-(chloromethyl)silatrane into the lungs increases its weight due to fibrogenesis of the connective tissue, growth of lymph nodes and an increase in the lipid and collagen contents of the lungs.

1-(Chloromethyl)silatrane does not produce any appreciable effect on the permeability of bilayered iipidic membranes. This preparation shows no mutagenic activity, as it has been found using strains of microorganisms (Salmonella typhimurium TA 1535 and TA 1538) which are very sensitive to the action of mutagens.

The simplest silatranes, $XSi(OCH_2CH_2)_3N$ where $X = H, C_6H_5, ClCH_2$ as well as 1-(chloromethyl)-3,7,10-trimethylsilatrane display no radio-protective effect on mice suffering from radiation sickness induced by 1000 r irradiation.

13 Conclusion

All the results of this survey show that silatranes are a new class of biologically active compounds having a broad spectrum of action. They will undoubtedly find a wide application in therapy, agriculture, fur-farming, poultry breeding, plant growing, cultivation of useful insects and microorganisms.

Acknowledgements. The author wishes to express his profound gratitude to his co-workers and colleagues: to the chemists V. M. Dyakov, G. I. Zelchan, V. P. Baryshok, E. J. Lukevits, M. S. Sorokin, L. I. Gubanova, G. A. Kuznetsova, N. M. Kudyakov, O. N. Florensova, Yu. A. Lukina, N. V. Semenova, to the biologists A. T. Platonova, L. A. Mansurova, L. T. Moskvitina, E. V. Bakhareva, L. V. Orgilyanova, K. M. Katrush, K. Z. Gamburg, R. I. Salganik, P. N. Platonova, V. V. Sadakh, to the pharmacologists I. G. Kuznetsov, G. A. Grigalinovich, J. J. Baltkais, E. A. Rudzit, A. I. Terekhina, to the agrochemists U. N. Madraimov, T. Ya. Tikhomirova, to the physicians I. I. Kolgunenko, N. S. Samoshkina, V. P. Sergeev, Yu. K. Skripkin, V. I. Kulagin, M. N. Nikitina, I. E. Kuznetsova, L. M. Tartakovskaya, to the oncologists K. P. Balitsky, I. G. Veksler, V. V. Vinitsky, A. A. Zidermane, to the entomologists N. N. Sinitsky, I. V. Vititnev, V. P. Drozda, N. G. Shkaruba, A. A. Rakitin, to the microbiologists E. Ya. Vinogradov, T. N. Semushina and other specialists whose enthusiastic participation in these studies has enabled all the data on the biological activity of silatranes to be interpreted and summarized.

References

1. Baltkais, J. J., Voronkov, M. G., Zelchan, G. I.: Izv. AN Latv. SSR, ser. khim. *1964*, N 2, 102; C. A. *61*, 9932 (1964)
2. Voronkov, M. G., Zelchan, G. I., Lukevits, E. J.: Silicon and life, Riga: Zinatne publishing house 1971; 2nd ed. 1977; Silicium und Leben, Akademie-Verlag, 1975
3. Voronkov, M. G., Zelchan, G. I.: Khimia geterotsikl. soed. *1965*, 51
4. Voronkov, M. G.: Pure Appl. Chem. *13*, 35 (1966)
5. Voronkov, M. G.: Vestn. AN SSSR *38*, N 10, 48 (1968)
6. Voronkov, M. G., Seltschan, G. I., Lapsina, A., Pestunowitsch, W. A.: Z. Chem. *8*, 214 (1968)
7. Voronkov, M. G.: Pure Appl. Chem. *19*, 399 (1969)
8. Voronkov, M. G., Lukevits, E. J.: Usp. khim. *30*, 2173 (1969)
9. Voronkov, M. G.: Vestn. AN SSSR *41*, N 11, 58 (1972)
10. Voronkov, M. G.: Chem. Brit. *9*, 411 (1973)
11. Voronkov, M. G.: Kagaku *28*, 213 (1973)
12. Voronkov, M. G.: Ceskoslovenska Farmacie *22*, 406 (1973)
13. Voronkov, M. G.: XXIVth Intern. Congress of Pure Appl. Chem., Vol. 4, p. 45–66. London: Butterworth 1974
14. Voronkov, M. G.: Ann. Reports in Medicinal Chem. *10*, 265 (1975)
15. Voronkov, M. G.: Jeszcze jeden, pierwiastek zycia, Politechnika Gdanska, 1975, p. 31
16. Voronkov, M. G.: Problemy, *23*, 667 (1967); Khimia i Zhizn, N 6, 69 (1966)
17. Zelchan, G. I.: Candidate dissertation, Institute of Organic Synthesis, AN Latv. SSR, Riga, 1967
18. Lukevits, E. J., Libert, L. I., Voronkov, M. G.: Usp. khim. *39*, 2005 (1970)
19. Radecki, A., Lukasiak, J., Ganowiak, Z., Vogel, S.: IVth Intern. Symposium on Organosilicon Chemistry, Vol. I, part 2, p. 12. Moscow 1975; Bromat. Chem. Toksykol. *9*, N 4, 495 (1976); Radecki, A., Lukasiak, J., Jamrogiewic, Z., Wrzesniowska, K.: Acta Pol. Pharm. *33*, 177 (1976)
20. Voronkov, M. G., Dyakov, V. M., Baryshok, V. P.: Zh. obshch. khim. *43*, 444 (1973)
21. Voronkov, M. G., Dyakov, V. M., Baryshok, V. P.: Zh. obshch. khim. *45*, 1650 (1975)
22. Voronkov, M. G., Dyakov, V. M., Baryshok, V. P., Tandura, S. N., Mironov, V. F.: Zh. obshch. khim. *45*, 1902 (1975)
23. Voronkov, M. G., Dyakov, V. M., Baryshok, V. P.: Zh. obshch. khim. *47*, 797 (1977)
24. Voronkov, M. G., Dyakov, V. M., Lukina, Yu. A., Samsonova, G. A., Kudyakov, N. M.: Zh. obshch. khim. *45*, 2010 (1975)
25. Voronkov, M. G., Dyakov, V. M., Samsonova, G. A., Lukina, Yu. L., Kudyakov, N. M.: Izv. AN SSSR, ser. khim. *1974*, 2794
26. Voronkov, M. G., Dyakov, V. M., Gubanova, L. I.: Izv. AN SSSR, ser. khim. *1974*, 657
27. Voronkov, M. G., Dyakov, V. M., Florensova, O. N.: Zh. obshch. khim. *45*, 1902 (1975)
28. Voronkov, M. G.: Sb. "Issledovania v oblasti fiziki i khimii kauchukov i resin", L., 1975, p. 265
29. Voronkov, M. G., Zelchan, G. I.: IX Mendeleev s'ezd po obshch. i prikl. khimii, M., 1965, p. 10.
30. Beiter, C. B., Schwarcz, M., Grabtree, G.: Soap Chem. Spec. *46*, 38 (1970)
31. Beiter, C. B., Schwarcz, M., Grabtree, G.: 5-p-Chlorophenylsilatrane, a new single-dose rodenticide, M & T Chemicals Inc., Bulletin, 8 pp., Rahway, New Jersey (1971); C. A. *73*, 120748 (1970)
32. Greaves, J. H., Redfern, R., Tinoworth, H.: J. Hyg. *73*, N 1, 39 (1974)
33. Bien, E.: Pharmazie *26*, 224 (1971)
34. Bien, E.: Pharmazie *26*, 577 (1971)
35. Garson, L. R., Kirchner, L. K.: J. Pharm. Sci. *60*, N 8, 1113 (1971)
36. Hulce, V. D., Rech, R. H.: Fed. Proc. *33*, 510 (1974)
37. Hulce, V. D., Rech, R. H.: Organosilicon Award Symposium, St. Louis, 1974

37a. Lukevics, E. J., Zelchan, G. I., Barton, T. J., Lapsina, A. F., Yudeika, I. A.: Izv. Akad. nauk. Latv. SSR, ser. khim. *1978*, 747

38. Atare, Z., Voronkov, M. G., Gutberg, S., Zelchan, G., Zile, A., Kimenis, A., Kruzmetra, L., Lukevits, E.: Kremneorganicheskie soed., Tr. Soveshch., M., vyp. IV, 1966, p. 25
39. Zelchan, G. I., Voronkov, M. G.: Rezultaty nauchno-issled. rabot po sozdaniyu novykh pestitsidov, vnedreniyu ikh v proizvodstvo i primeneniyu v selskom khozaistve, sektsia I. Izyskanie pestitsidov. Tezisy dokl. i soobshch. vsesoyuz. konf., M., 1972, p. 3
40. Zelcans, G., Lukevics, E.: Silicijs dzivaja daba, Riga, 1976
41. Shevchenko, S. S., Kuznetsov, I. G., Moskvitina, L. T., Pukhnarevich, V. B., Dyakov, V. M., Platonova, A. T.: Tezisy dokl. 5 konf. molodykh uchenykh, Zinatne publishing house, Riga, 1976, p. 10
42. Moskvitina, L. T., Kononenko, V. F., Platonova, A. T., Dyakov, V. M., Florensova, O. N., Tataurova, G. P., Voronkov, M. G.: Tezosy dokl. I Vsesoyuz, simpoziuma "Biol. akt. soed. elementov IV B gruppy", Irkutsk, 1975, p. 45
43. Voronkov, M. G., Platonova, A. T., Kuznetsov, I. G., Shevchenko, S. G., Meierova, E. A., Suslova, S. K., Emelyanov, I. S., Dyakov, V. M., Ustinova, N. G., Zelchan, G. I., Lukevits, E. J.: Izv. AN Latv. SSR, ser. khim., 1977, p. 204
43a. Voronkov, M. G., Zelchan, G. I., Chernyshev, E. A., Tabenko, B. M., Savushkina, V. I.: USSR 364621 (1973); C. A. *78*, 159845 (1973); khim. Geterotsikl. soed. *1976*, 772
44. Casida, J. E., Eto, M., Moscioni, A. D., Engel, J. L., Milbrath, D. S., Vercade, J. G.: Toxicol. appl. pharmacol. *36*, 261 (1976)
44a. Brandt, K., Mattson, H., Heilbronn, E.: Acta Pharm. Toxicol. *41*, 42 (1977), Suppl. 4
44b. Mattson, H., Brandt, K., Heilbronn, E.: Sixth Internat. Meeting of the Internat. Soc. for Neurochem. Abstracts, Copenhagen, 1977, p. 56
45. Voronkov, M. G., Emelyanov, I. S., Zelchan, G. I., Dyakov, V. M., Kuznetsov, I. G.: Khimia geterotsykl. soed. *1975*, 35
46. Voronkov, M. G., Sorokin, M. S., Kletsko, F. P., Dyakov, V. M., Vlasova, N. N., Tandura, S. N.: Zh. obshch. khim. *45*, 1395 (1975)
47. Voronkov, M. G., Sorokin, M. S., Dyakov, V. M., Kletsko, F. P., Vlasova, N. N.: Zh. obshch. khim. *45*, 1649 (1975)
48. Voronkov, M. G., Vlasova, N. N., Platonova, A. T., Kletsko, F. P., Tsykhanskaya, I. I., Kuznetsov, I. G., Shevchenko, S. G., Sadakh, V. V., Suslova, S. K.: Dokl. Akad. Nauk SSSR *229*, 1011 (1976)
49. Voronkov, M. G., Sorokin, M. S., Dyakov, V. M.: Zh. obshch. khim. *45*, 1394 (1975)
50. Voronkov, M. G., Dyakov, V. M., Florensova, O. N., Baryshok, V. P., Kuznetsov, I. G., Chvalovsky, V.: Coll. *42*, 480 (1977)
51. Bellet, E. M., Casida, J. E.: Science *182*, 1135 (1973)
52. Franke, Z.: Khimia otravlyayushchikh veshchestv, "Khimia", M., 1973, I, p. 396
53. Kononenko, V. F., Moskvitina, L. T., Platonova, A. T., Kuznetsov, I. G., Dyakov, V. M., Voronkov, M. G.: Tesisy dokl. I Vsesoyuzn. simpoziuma "Biol. aktivn. soed. elementov IVB gruppy, Irkutsk, 1975, p. 17
54. Kononenko, V. F., Minkina, O. I., Moskvitina, L. T., Platonova, A. T., Kuznetsov, I. G., Voronkov, M. G.: Sb. dokl. II Vsesoyuzn. simpoziuma "Biol. aktivn. soed. elementov IVB gruppy, Irkutsk, 1977, p. 110
55. Shevchenko, S. G., Platonova, A. T., Sadakh, V. V., Voronkov, M. G.: Ref. 54, p. 95
56. Kazimirovskaya, V. B., Platonova, A. T., Pestunovich, A. E., Vlasova, N. N., Vasyanovich, G. P., Dyakov, V. M., Baryshok, V. P., Voronkov, M. G.: Ref. 54, p. 99
57. Pinigina, N. M., Platonova, A. T., Lukina, Yu. L., Stankevich, V. N., Baryshok, V. P., Tsykhanskaya, I. M., Dyakov, V. M., Voronkov, M. G.: Ref. 54, p. 84
58. Ardavichene, T. A., Pinigina, N. M., Lukina, Yu. A., Platonova, A. T., Voronkov, M. G.: Ref. 54, p. 89
59. Terekhina, A. I., Gritsina, G. I., Kaverina, L. P., Lisitsa, L. I., Mikhailovskaya, L. L.: Ref. 53, p. 17
60. Mansurova, L. A., Platonova, A. T., Kuznetsov, I. G., Voronkov, M. G.: Ref. 53, p. 28
61. Mansurova, L. A., Kuznetsova, I. G., Platonova, A. T., Voronkov, M. G.: Ref. 53, p. 12
62. Kuznetsov, I. G., Voronkov, M. G., Dyakov, V. M., Platonova, A. T.: IVth Intern. Symp. Organosilicon Chem., Abstract of Papers, Moscow, 1975, Vol. I. part I, p. 187

63. Grigalinovich, G. A., Atare, Z. A., Milyakova, I. V., Zelchan, G. I., Lukevits, E. J.: Ref. 54, p. 136
64. Slutsky, L. I., Symkhovich, B. Z., Dombrovska, L. E., Astapenok, E. B., Grigalinovich, G. A., Amelin, A. Z., Zelchan, G. I., Lukevits, E. J.: Ref. 54, p. 140
65. Mansurova, L. A., Kuznetsov, I. G., Platonova, A. T., Voronkov, M. G.: Ref. 54, p. 131
66. Simkhovich, B. Z., Zamaraeva, T. V., Zelchan, G. I., Lukevits, E. J., Mazurov, V. I.: Biokhimia *42,* N 6 (1977), 1128
67. Kuznetsov, I. G., Platonova, A. T., Suslova, S. K., Dyakov, V. M., Voronkov, M. G.: Ref. 53, p. 34
68. Sorokin, M. S., Kuznetsov, I. G., Dyakov, V. M., Suslova, S. K., Pushechkina, T. A., Voronkov, M. G.: Ref. 54, p. 17
68a. Voronkov, M. G., Platonova, A. T., Mansurova, L. A., Kuznetsov, I. G., Zelchan, G. I., Dyakov, V. M.: Brit. 1465455 (1975); Fr. 2321295 (1977); Ger. Offen. 2530255 (1977); C. A. *86,* 54101 (1977)
69. Bakhareva, E. V., Platonova, A. T., Kuznetsov, I. G., Voronkov, M. G.: Ref. 53, p. 30
69a. Bakhareva, E. V., Martynyuk, N. I., Platonova, A. T., Voronkov, M. G.: Ref. 53, p. 10
70. Grigalinovich, G. A., Atare, Z. A., Milyakova, N. V., Zelchan, G. I., Lukevits, E. J.: Ref. 54, p. 113
71. Kuznetsova, I. P., Platonova, A. T., Katkevich, A. I., Bakhareva, E. V., Galchenko, L. I., Velichenskaya, T. B.: Ref. 54, p. 125
72. Voronkov, M. G., Platonova, A. T., Bakhareva, E. V., Kuznetsov, I. G., Dyakov, V. M., Seltschan, G. I., Kalgunenko, I. I.: Ger. Offen. 2615654 (1977)
73. Bakhareva, E. B., Platonova, A. T., Tartakovskaya, L. M., Voronkov, M. G.: Ref. 53, p. 14
74. Sergeev, V. I., Irlyanova, S. V.: Ref. 54, p. 117
75. Finogenov, N. N., Doctor dissertation, St. Petersburg, 1909
76. Bogomolets, A. A.: Rol'fiziologicheskoj sistemy soedinitelnoj tkani v yavleniakh immuniteta i neoplazmy, Izb. tr., 2, Kiev, 1957
77. Vasiliev, Yu. M.: Soedinitelnaya tkan i opukholevyj rost v eksperimente, M., 1961
78. Voronkov, M. G., Grigalinovich, G. A., Zelchan, G. I.: Dokl. Akad. Nauk SSSR *200,* 967 (1971)
79. Balitsky, K. P., Veksler, I. G., Dyakov, V. M., Kuznetsov, I. G., Voronkov, M. G.: Ref. 54, p. 153
80. Zelchan, G. I., Lapsina, A. F., Solomennikova, I. I., Dauvarte, A. Z., Zidermane, A. A., Lukevits, E. J.: Ref. 54, p. 28
80a. Lukevics, E., Lapsina, A. F., Zelchan, G. I., Dauvarte, A., Zidermane, A. A.: Izv. Akad. nauk Latv. SSR, ser. khim. *1978,* 338
80b. Lukevics, E., Khokhlova, L. N., Zidermane, A. A., Dauvarte, A.: USSR 579275 (1977)
81. Dyakov, V. M., Baryshok, V. P., Florensova, O. N., Voronkov, M. G., Kuznetsov, I. G., Moskvitina, L. T., Shevchenko, S. S.: Tezisy dokl. I Vsesoyuzn. konf. "Sintez i mekhanizm deistvia fiziol. aktivn. veshchestv, Odessa, 1976, p. 38
82. Shevchenko, S. S., Kuznetsov, I. G., Platonova, A. T., Voronkov, M. G.: Ref. 53, p. 33
83. Vinitsky, V. B., Shmalko, Yu. P., Balitsky, K. P., Voronkov, M. G.: Ref. 54, p. 74
84. Baryshok, V. P., Dyakov, V. M., Voronkov, M. G., Shevchenko, S. S., Kuznetsov, I. G.: Ref. 54, p. 14
85. Stepanenko, B. N., Raevsky, K. S., Kopkov, V. I., Pochatov, Yu. M., Luzin, A. P.: Ref. 54, p. 80
86. Voronkov, M. G., Platonova, A. T., Dyakov, V. M., Katrush, K. M., Kazakul, A. I., Kuznetsov, I. G., Mansurova, L. A.: USSR 541473 (1977); C.A., *87* (1977) 83547
87. Katrush, K. M., Voronkov, M. G., Dyakov, V. M.: Ref. 54, p. 162
88. Voronkov, M. G., Vititnev, I. V., Drozda, V. F., Dyakov, V. M., Sinitsky, N. N., Shkaruba, N. G.: USSR 531523 (1976); C. A., *86* (1976) 54101
89. Vititnev, I. V., Drozda, V. F., Shkaruba, I. G., Sinitsky, N. N., Orgilyanova, L. V.: Ref. 54, p. 170
90. Drozda, V. F., Vititnev, I. V., Shkaruba, N. G., Sinitsky, N. N., Skorobogatova, V. I., Vugmeister, E. K., Dyakov, V. M., Voronkov, M. G.: Ref. 54, p. 174

91. Shkaruba, N. G., Vititnev, I. V., Drozda, V. F., Sinitsky, N. N., Dyakov, V. M., Voronkov, M. G.: Ref. 54, p. 184
92. Rakitin, A. A., Novozhilov, K. V., Dyakov, V. M., Voronkov, M. G.: Ref. 54, p. 177
93. Voronkov, M. G., Yititnev, I. V., Drozda, V. F., Dyakov, V. M., Sinitsky, N. N., Shkaruba, N. G., Sorokin, M. S., Baryshok, V. P.: Dokl. Akad. Nauk SSR *239,* 238 (1978)
94. Orgilyanova, L. V., Osharova, L. M., Gamburg, K. Z., Dyakov, V. M., Semenova, N. V., Platonova, A. T., Reimers, P. E., Voronkov, M. G.: Ref. 53, p. 44
95. Semenova, N. V., Dyakov, V. M., Voronkov, M. G., Melkostupova, V. N.: Ref. 54, p. 58
96. Orgilyanova, L. V., Gamburg, K. V., Semenova, N. V., Dyakov, V. M., Voronkov, M. G.: Dokl. Akad. Nauk SSSR *227,* 1486 (1976)
97. Tikhomirova, V. Y., Dyakov, V. M., Voronkov, M. G.: Ref. 54, p. 189
98. Bakhareva, E. V., Kuznetsov, I. G., Platonova, A. T., Voronkov, M. G.: Ref. 54, p. 146
99. Tizenberg, G. M., Kuznetsov, I. G., Platonova, A. T., Dyakov, V. M., Voronkov, M. G.: Ref. 54, p. 150

Received November 2, 1978

Author Index Volumes 26–84

The volume numbers are printed in italics

Structure and Bonding

Editors: J. D. Dunitz, J. B. Goodenough, P. Hemmerich, J. A. Ibers, C. K. Jørgensen, J. B. Neilands, D. Reinen, R. J. P. Williams

Volume 33

New Concepts

1977. 85 figures, 47 tables. IV, 214 pages
ISBN 3-540-08269-7

Contents:

W. E. Wallace, S. G. Sankar, V. U. S. Rao: Crystal Field Effects in Rare-Earth Intermetallic Compounds
D. K. Hoffmann, R. Ruedenberg, J. G. Verkade: Molecular Orbital Bonding Concepts in Polyatomic Molecules: A Novel Pictorial Approach
K. D. Warren: Ligand Field Theory of *f*-Orbital Sandwich Complexes
K. Schubert: The Two-Correlations Model, a Valence Model for Metallic Phases
C. Linares, A. Louat, M. Blanchard: Rare-Earth Oxygen Bonding in the $LnMO_4$ Xenotime Structure: Spectroscopic Investigation and Comparative Study of Ligand Field Models

Volume 34

Novel Aspects

1978. 37 figures, 39 tables. IV, 220 pages
ISBN 3-540-08676-5

Contents:

S. Hubert, M. Hussonnois, R. Guillaumont: Measurement of Complexing Constants by Radiochemical Methods
C. K. Jørgensen: Predictable Quarkonium Chemistry
R. A. Bulman: Chemistry of Plutonium and the Transuranics in the Biosphere
J. W. Buchler, W. Kokisch, P. D. Smith: Cis, Trans, and Metal Effects in Transition Metal Porphyrins
D. K. Koppikar, P. V. Sivapullaiah, L. Ramakrishnan, S. Soundarajan: Complexes of the Lanthanides with Neutral Oxygen Donor Ligands

Volume 45

New Theoretical Aspects

1978. 41 figures, 15 tables. IV, 176 pages
ISBN 3-540-08887-3

Contents:

J,-F. Labarre: Conformational Analysis in Inorganic Chemistry: Semi-Empirical Quantum Calculations vs. Experiment
D. B. Cook: The Approcimate Calculation of Molecular Electronic Structures as a Theory of Valence
D. W. Smith: Applications of the Angular Overlap Model
C. Furlani, C. Cauletti: He(I) Photoelectron Spectra of *d*-Metal Compounds

Volume 36

Inorganic Chemistry and Spectroscopy

1979. 69 figures, 22 tables. IV, 178 pages
ISBN 3-540-09201-3

Contents:

R. J. H. Clark, B. Stewart: The Resonance Raman Effect. Review of the Theory and of Applications in Inorganic Chemistry
A. Simon: Structure and Bonding with Alkali Metal Suboxices
C. Daul, C. W. Schläpfer, A. von Zelewsky: The Electronic Structure of Cobalt (II) Complexes with Schiff Bases and Related Ligands

Volume 38

Structural Problems

1979. 40 figures. 43 tables. IV, 216 pages
ISBN 3-540-09455-5

Contents:

D. Reinen, C. Friebel: Local and Cooperative Jahn-Teller Interactions in Model Structures, Spectroscopic and Structural Evidence
E. E. Hellner: The Frameworks (Bauverbände) of the Cubic Structure Types
J. Shamir: Polyhalogen Cations

Volume 38

G. A. Somorjai, M. A. van Hove

Adsorbed Monolayers on Solid Surfaces

1979. 31 figures, 9 tables. Approx. 250 pages
ISBN 3-540-09582-9

Contents:

Introduction. – Principles of Monolayer Adsorption. – Principles of Ordering of Adsorbed Monolayers. – Methods of Structure Analysis. – Overview of the Ordering of Adsorbates. – Surface Crystallography of Ordered Monolayers of Atoms. – Surface Crystallography of Ordered Multi-Atomic and Molecular Monolayers.

Springer-Verlag
Berlin Heidelberg New York

GPSR Compliance

The European Union's (EU) General Product Safety Regulation (GPSR) is a set of rules that requires consumer products to be safe and our obligations to ensure this.

If you have any concerns about our products, you can contact us on ProductSafety@springernature.com

In case Publisher is established outside the EU, the EU authorized representative is:

Springer Nature Customer Service Center GmbH
Europaplatz 3
69115 Heidelberg, Germany

Batch number: 10620881

Printed by Printforce, the Netherlands